RECENT ADVANCES IN POMEGRANATE

Dr. M.K. Sheikh
M.Sc. (Agri.), Ph.D., FNESA
Professor and Head Department of Horticulture
College of Agriculture
Bijapur

Dr. Manjula. N.
M.Sc. (Agri.), Ph.D., FNESA
Professor and Head Department of Agril. Extension
College of Agriculture
Bijapur

ibdc publishers
PUBLISHING GREEN

Published by

ibdc publishers
PUBLISHING GREEN

Khushnuma Complex Basement
7, Meerabai Marg (Behind Jawahar Bhawan)
Lucknow 226 001 U.P. (INDIA)
Tel. : 91-522-2209542, 2209543, 2209544, 2209545
Fax : 0522-4045308
E-Mail : ibdco@airtelmail.in

First Edition 2013

ISBN 978-81-8189-194-5

Composed & Designed at :
Panacea Computers
3rd Floor, Agrawal Sabha Bhawan
Subhash Mohal, Sadar Cantt. Lucknow-226 002
Phone : 0522-2483312, 9335927082, 9452295008
E-mail : prasgupt@rediffmail.com

Printed at:
Salasar Imaging Systems
C-7/5, Lawrence Road Industrial Area
Delhi - 110 035
Tel. : 011-27185653, 9810064311

Dr. H.P. SINGH
Deputy Director General
(Horticulture)

भाकृअनुप
ICAR

Indian Council of Agricultural Research
Krishi Anusandhan Bhavan-II,
Pusa, New Delhi-110012
Ph. : (Off.) 2584 2068
(Resi.) 25841189
(Mob.) 9871450730
Fax : 91-11-2584 1976
E-mail : hpsingh@icar.org.in

Foreword

Pomegranate, although found in India in wild form, had its early domestication in Mediterranean region. Currently, the crop is becoming important both in tropical and sub-tropical regions of the globe. In India, pomegranate has expanded well in dry land and its providing unprecedented return per unit area, especially in Maharashtra, which contributes 70 per cent of the production. The pomegranate fruit is rich in potentially active physiochemicals and targets in reduction of risk for a number of diseases. In the health conscious population, the consumption has increased in different countries. Appreciable increase in export from India is also seen in recent years to UAE and Europe. The increasing demand of pomegranate in the international market has widened the scope of production and trade. In the last decades, there has been a pioneering work to develop new cultivars and technologies under the National Agriculture Research System. The information pertaining to pomegranate technologies is scattered. The publication "Recent Advances in Pomegranate" covers a great deal of information on various aspects of pomegranate addressing to increase the productivity.

I compliment the authors for compiling valuable information together on pomegranate cultivation. I am sure the book will be of great use to all who are concerned with pomegranate.

(H.P. Singh)

Preface

Pomegranate (*Punica granatum* L) belongs to the family punicaceae is native of Iran. In India it is grown over an area of 1.13 lakh hectares with production of 7.92 lakh tonnes. Ganesh was popular variety till recently, which has been replaced with Bhagwa which is export oriented variety grown by majority of farmers. Although India rank 2nd in fruit production but its share in world trade is negligible for fresh fruits. During 2007-08 we could export 0.035 million tones indicating bright future for export, our destinations of export are United Arab Emirates (UAE), United Kingdom, Netherland.

Pomegranate fruit having rich medicinal value because of Antioxidant properties, sugars, vitamins and minerals which are helpful in controlling heart diseases, cancer, diabetes, Aids etc.

Pomegranate is hardy crop adopted to wide range of agro climatic conditions and is grown in poor soil, yet respond well to application of nutrients and water. There are good number of cultivars like Bhagwa, Ruby, Ganesh in India, wonderful in USA and Israel, Hicaznar in Turkey, Kandkhar in Afghanistan, Shahavar (Saveh) in Iran, Dolce, Alappi in Italy.

New storage and processing technology and innovative packing techniques have made the pomegranate to reach the distant markets. The value added products like pomegranate juice, processed products like wine, squash, syrups, jam, jelly, candy and anardana are gaining momentum in world trade.

Recently the bacterial blight has become a threat for pomegranate cultivation, since no Indian varieties are resistant to this bacterial blight. However, it advised to take up Hasta bahar crop by following good management practices.

Among Abiotic stress fruit cracking and splitting, internal break down of aril and sun scald are important in dry parts of the globe with strong sun rays. This can be reduced by covering the fruits with paper or thin guny bags. Oflate *Pseudomonas fluorescens* (10^9 cfu/g) plus IBA 1500 ppm helps in good rooting of the plant.

Induction of flowers in Hasta bahar is problematic because of improper rest, which leads to vigorous growth, application of lihocin 500 ppm, 750 ppm, 1000 ppm with pinching of new shoot tips and application of bio stimulant help in flowering. Pruning the tip of spur help in induction of Hermophrodite flowers as genetically pomegranate bear flower terminally, occasionally axillary.

The authors are thankful to Mr. Mazagi for good typing and assistance extended by Mr. S. R. Mulla & Md.Bilal.Sheikh are solicited.

Dr. M. K. Sheikh
and
Dr. Manjula. N

Contents

Chapter 1
Introduction

POMEGRANATE (*Punica granatum* L.)

Pomegranate belongs to the family punicaceae. The flowers are dioecious actinomorphous, single or arranged in few flowered clusters (similar to reduced dichasium), sepals are grown together. The calyx is leathery like and coloured. The petals are bright coloured. The stamens are numerous. The ovary is inferior, adherent with the receptacle to the middle. The loculi are organized in 1-2 (rarely) 3 circles. The style is single, the stigma is capitate and the pomegranate fruit is having leathery pericarp, seeds and numerous with juicy sarcotesta. Cotyledons are wrapped with no endosperm. The family includes two monotypic genera socotria. It always flower terminally and occasionally axilllary.

Distribution

Pomegranate is grown around the world within the area lying between latitude 41 North and 42 South. It is important to distinguish between Northern boundary of the commercial culture of pomegranate ,the Northern boundary of its cultivation as a fruit culture grown in open soil as decorative plant.

The Pomegranate (*Punica granatum* L.) is one of the ancient and highly praised favourite fruits. It is commercially grown, apart from India, in a number of countries for its sweet-acidic fruits, which provide a cool refreshing juice, and is valued for its medicinal properties. Its popularity is also due to the ornamental nature of the plant which bears bright red, attractive flowers. To highlight its importance, it was chosen as a symbol of the 18th International Horticultural Congress held during 1970, showing pomegranate in a basket.

Pomegranate can successfully be cultivated even under conditions of severe drought and frost. The fruit rind, juice, leaf and roots are used in the preparation of various Ayurevedic medicines. The Juice and seeds also contain large quantities of tannin and agollic acid, which is essential in curing several diseases.

NUTRITIVE VALUE

(*Per 100 gm of edible portion*)

Moisture	-	78.00 gms.
Protein	-	1.60gms.
Fat	-	0.10 gms.
Minerals	-	5.10 gms.
Phosphorous	-	0.07 gms.
Iron	-	0.30 mg.
Riboflavin	-	100.00 mg.
Vitamin 'C'	-	16.00 mg.

In the recent past pomegranate cultivation gained momentum due to the production technologies developed by the scientists. Among the different states Maharashtra is leading in pomegranate production. It is also grown commercially In Karnataka, Andhra Pradesh, Rajasthan, Gujarat and Tamil Nadu. As there is tremendous scope for exports, the central and state governments are providing several subsidies. Subsidies are being provided in several states for the installation of drip irrigation in pomegranate orchards, and for the control of bacterial blight . For competing in International trade our cultivators have to produce pomegranate fruits of International quality and standards. To achieve these standards the cultivation details are furnished hereunder.

Pomegranate (*Punica granatum* L.) belongs to the natural order *Punicaceae. Punica* perhaps is the only known genus of this family, which includes large shrubs over small trees with 2 species. *Punica Protopunica* Balf. S is found wild in Socotra Island and the other *Punica granatum* is cultivated in tropical and sub-tropical parts of the world. Punica granatum has been classified into two sub-species *Chlorocarpa* and *Porphyrocarpa,* each having 2 varieties. These sub-species have been established on the basis of the colour of the ovary, a stable feature which is retained when they are reproduced by seeds. Sub-species *Chlorocarpa* is mainly found in the *Trancaucasus,* whereas, the second sub-species *Porphyrocarpa* is mainly distributed in central Asia.

Pomegranate (*Punica granatum*) one of the favourite table fruits of tropical and sub-tropical regions. The fruit is grown in Iran and is extensively cultivated in Mediterranean countries like Spain, Morocco, Egypt, Iran, Afghanistan ,Turkey and Baluchistan. It is also grown to some extent in Burma, China, Japan, USA (California), USSR and India. It is also found growing in Bulgaria and in Southern Italy.

Pomegranate (*Punica granatum*) is native to Iran and surrounding area. Pomegranate is associated with the most ancient civilization in the Middle East and later it naturalized over the whole Mediterranean region.

It flourished particularly well in Spain and the city of Granada where high quality pomegranate are grown.

Pomegranate is quite suitable to hot and arid regions and therefore it also spread to India and China. Spänish missionaries brought the pomegranate to the new world including Mexico and California in U.S.A.

India is the World's leading producing country of pomegranates. Approximately 1,13,000 hectares area is cultivated for pomegranate. The details of area and production in different states are given below.

Table-1: Pomegranate production in India

Sl. No.	State	Area under Cultivation (ha)	Production in tons	Productivity Tons/ha
1	Maharashtra	88500	568200	10.0
2	Karnataka	12700	124900	5.80
3	Gujarat	41000	38400	7.40
4	Andhra Pradesh	58000	47800	8.00
5	Rajasthan	4000	17000	11.30
	Total	112500	792500	8.60

Its cultivated area is as much as 1,12,500 ha which are under cultivation in India with production of 7,92,500 tonnes. However, at present, Maharashtra is the,primary pomegranate producing state. Area under pomegranate in this state has increased from 31,549 ha in 1996-97 to 88,500 ha in 2004-05.

Next to Maharashtra, maximum pomegranate cultivation takes place in Karnataka where the area has increased from 7,584 ha in 1996-97 to 12,700 in 2004-05.

Pomegranate is also cultivated in other parts of the world. However, India remains the leading producing country (7,92,000 tons). Other countries well-known for pomegranate production are Spain (1,00,000 tons approx.) and Iran (8,00,000 tons approx).

Pomegranate is also quite a nutritive fruit as it is rich in riboflavin, niacin, and ascorbic acid and also in potassium.

In India, it is largely used as desert fruit and its fresh juice is much relished. In foreign countries, it is commonly used for decorative purposes its fresh kernels are used for garnishing desserts and salads. Pomegranate juice is mainly used for making Grenadine (a type of syrup). Which is further used for flavouring of mixed drinks, ice creams or desserts.

Edible quality of Indian Pomegranates especially of soft seeded ones, i.e., Ganesh, G-137, Mridula, Jalore seedless, Ruby, seems to be much better, as their T.S.S. varies from 15 to 17 Degree Brix and acids between 0.3% to 0.45% whereas Spanish Pomegranate have T.S.S. between 17 to 18 Degree Brix and acids from 1.5% to 1.8%. Probably this is the reason why Spanish Pomegranate are more used for decoration, garnishing or for syrup making, whereas Indian Pomegranates are more used for fresh fruit juice purposes.

Concentrated pockets of Pomegranate in India

Table 2: Concentrated pockets of pomegranate growing areas

State	District	Productive area in ha	Production in tons
Maharashtra	Solapur	15,157	1,51,570
	Nasik	10,800	1,08,000
	Sangli	5,300	53,000
	Ahmednagar	2,615	26,150
Karnataka	Bijapur	8,454	49,059
Andhra Pradesh	Anantpur	2,000	16,000

Maharashtra state accounts for 78% of the total area in India and 84% of the total production in the country. Pomegranate is concentrated in 4-5 districts and prominent districts along with their area under pomegranate are as follows:

The prominent pockets where pomegranate area are concentrated are Solapur, Nasik, Sangli, and Ahmednagar in Maharashtra. Bijapur, Bagalkot and Koppal in Karnataka state. In Andhra Pradesh also concentrated area for Pomegranate cultivation has emerged in Anantpur district.

Maximum area of Pomegranate is occupied by Ganesh variety (approx. 90%) and other important varieties include G-137, Mirdula, Bhagwa and Arkata. All these cultivars are soft seeded having high T.S.S and low acids with light pink coloured arils to dark red coloured arils.

Export scenario of fresh Pomegranate from India

Detailed data on exports of fresh pomegranates from India are presented in Table 3. This shows India plays a significant role in the global market. Recently exports from India of Pomegranates have extended to several countries. The figures indicate that during 1999-2000 as much as 5,726 tons of pomegranates were exported for a total of Rs. 11.5 crores. However, during 2000-2002, exports declined to 4,455 tons for a total of Rs. 9.91 crores.

Most notable point is that the exports for the last four years are almost varying between 4,000 tons to 5,500 tons. These exports form only approx. 1.0% of the total production whereas Spain produces approx. 1,00,000 tons of pomegranate and exports 75% of the production.

A critical analysis of export figures show that during 2000-2001, maximum exports of pomegranate were Gulf countries and particularly UAE (2,922 tons), Saudi Arabia (170 tons), Bahrain (193 tons) and Kuwait (65 tons).

During 1999-2000 exports of pomegranate to Middle East countries were much higher i.e., 3,159 tons to UAE; 1,232 tons to Saudi Arabia, 160 tons to Bahrain and 139 tons to Kuwait. This suggests that exports to this region can be considerably accelerated as there seems to be excellent potential.

As far as exports of fresh pomegranate to SAARC countries are concerned, sizeable quality is exported to Bangladesh, Sri Lanka and Pakistan. In these countries also there is a good scope for stepping up exports of fresh pomegranate.

After Gulf and SAARC countries, next in terms of quantum of exports are the European countries. U.K. is the principal importer buying 161 tons in 2000-2001. However, in earlier years, U.K. imported in much larger quantities, i.e.., 358 tons in 1997-98; 273 tons in 1998-99 and 241 tons in 1998-99. Similarly, countries like France, Germany and Switzerland imported 35 tons, 48 tons and 21 tons respectively during 1998-99. More recently in 2000-2001, our exports to Europe have diversified as we exported 140 tons to Italy, 80 tons to Netherlands and 31 tons to Belgium. This shows there is wide acceptability for Indian pomegranate and this needs to be boosted by nurturing the markets.

There is renewed interest in exploration to Far East countries where there is a good potential for marketing. Pomegranate was exported to the Phillipines (35 tons in 1998-99), Singapore (22 tons in 1999-2000) and Hong Kong (15 tons in 1998-99).

There are very little scope for exports to African countries except in South Africa, where 47 tons of fresh pomegranates were exported during 2000-2001. This needs to be further viewed in terms of marketing it in future. However, there has been a negligible export to other countries in this region.

Regarding North America, there is scope for exports to Canada. During 1999-2000, 60 tons of fresh pomegranates were exported and during 2000-2001, tons, 51 tons were exported. Price realization is also quite high and considering longer shelf life of pomegranates, countries in this continent like Brazil, Mexico, Trinidad,

Barbados and Bahamas where small quantities of fresh pomegranates have been exported, efforts are needed to expand the market there.

World market for pomegranates

Production

Data on production of pomegranate is not available in F.A.O. statistics. However, old data on website, shows that the present production of pomegranate is of the order of 10,00,000 tons.

Out of one million tons production, India is the leading producer of pomegranate. With a production of 7,92,000 tons, Spain and Iran produce approx, 8,00,000 tons each. Rest of the produce comes form Egypt, Israel, Jordan, Lebanon, Tunisia, Pakistan, Thailand, Chile, Peru and U.S.A.

World Trade

As per estimates, total world trade is around 1,00,000 tons to 1,20,000 tons per year. However, no official statistics are available from F.A.O. or any other source. Out of this quantity, Spain alone exports about 70,000 to 75,000 tons per year. Spain is a dominant player in world exports. Next to Spain, Iran exports approx 15,000 tons, whereas India exports only 5,000 tons per year. Pomegranate from India and Iran are almost available throughout the year while Spanish pomegranates are in the peak in the months of September to December. Spain continues to dominate EU market due to advantage in freight cost over Iran and India. Iran also exports to Europe during November to March, However in Middle East market both Iran and India are the main exporters.

Important aspects of the pomegranate trade are that Spain produces only about 10% of the world's production but has more than 50% share in world exports. Whereas Indian produces almost 50% of the world pomegranates but has a mere 5% share in world exports.

a. Importing countries

The main importing countries are in Europe and Middle East. Imports of the order of 80,000 to 90,000 tons/year are done by EU market, comprising of U.K. France, Germany. The Netherlands, Sweden, Switzerland, Denmark, Belgium and Italy. Whereas Gulf countries import to the tune of 15,000 tons to 25,000 tons/ years. The major Middle East importing countries are Saudi Arabia, U.A.E. Bahrain, Kuwait, Qatar, Oman, etc.

b. Quality requirements of importing countries

The desirable fruit characters for export purposes are detailed below:

i Dark rose pink colour of fruit.

ii. Fruit weight of around 200-400 gm.
iii. Round shape of fruit.
iv. Uniform size and shape of the fruit in pack or box.
v. Dark rose pink arils.
vi. Softness of the seeds.
vii. Higher sugar contents of about 16 °B to 17 °B.
viii. Freedom from scars, russeting, disease spots, insect injury, and scratches.
ix. Smooth cutting at the stem end.
x. Bracts/calyx without any damage and having freshness.
xi. Pleasant flavour and aroma.

The wild pomegranate fruit is filled with angular hard seeds covered with a juicy, pink or yellowish white, sweet astringent acid pulp. The fruit is used by sun-drying the seeds along with the pulp. Which constitute the product 'anardana' a valuable commodity used as a condiment and for other culinary purposes. Methods have been standardized for the commercial exploitation of anardana. The seeds have oestrogenic activity due to the presence of 1.00 mg oestrogene and 0.036 mg coumestrol (a non-steroidal oestrogen), 100 g-1 of seeds *etc.* The products of pomegranate such as bottled juice, syrup and jelly are highly appreciated because of their nutritive and therapeutic qualities. The surface colour of fruit varies in few commercial cultivars from yellow with a crimson, from brownish red to bright red. The edible portion is the bright-red pulp surrounding the individual seeds, six anthocyanin pigments are responsible for the red color of pomegranate and identified as delphinidin-3-5-glucoside and 3,5-diglucosi de, cyaniding-3-glucoside, pelargonidin-3-glucoside and 3,5-diglucoside. Generally, there was an increase in juice pigmentation with fruit ripening. The concentration of pigments in juice obtained from matured fruits range from 50 to 100 mg anthocyanin and g-1 fresh weight of aril. The total amount of pigment in the juice was generally less in fruits with reddish skin than in those with yellow skin.

Asem, (2006) reported Pomegranate plantations are under irrigation. Recently the National Centre for Agricultural Research and Extension, Jordon implemented a base line survey as the first well documented applied study; the study resulted in the identification of sixteen varieties and according to the study results; 50% of them were sweet, 23% Japan (Sweet-Sour) and 18% sour. The highest productivity approximately was eight tones per donum. Most orchards belong to small and medium size holders. Fruit splitting and fruit rot are the main causes of product losses. Three promising national projects have been implemented to develop this sector: (i) Pomegranate national campaign to promote and marketing (ii) Jordanian pomegranate project and (iii) development of local community's project.

Yahya, (2009) says Iran is leading producer and exporter of pomegranate in the world that has a long history of cultivation. Although the proper climatic conditions for producing high quality and quantity pomegranate are known, most of the planting sites were not selected precisely. Many parts of Iran encounter rainfall limitation, in tolerable heat and severe cold in different seasons that may possibly cause different chilling or heat injuries. Therefore recognizing the most favorable regions for this crop is a high priority. In this study, ecological conditions, geographical location and physiography of studied area was firstly examined, and remote sensing techniques have made proper use of land, geographical information system was used to facilitate identification of suitable pomegranate planting sites. Based on GIS data, chilling probability layers, humidity index, day degree and rain distribution layers, the east part of Iran was divided into four parts including very desirable, desirable and unsuitable for planting pomegranate. The regions that showed the highest accordance between climatic parameters and pomegranate requirements are recommended for extending pomegranate cultivation sites in order to prevent climatic injuries.

Mozhgan *et. al.,* (2009) reported on pomegranate usage in the arid urban landscape says that, introducing landscape in arid place is expensive and lack of water is a serious problem in this area. Best way is maintenance of urban landscape where native plant can with stand hard situation. Pomegranate is an attractive shrub of small trees grows to 20 or 30 ft height. The plant is much branched, more or less spiny, and extremely long-lived, some specimens at Versailles known to have survived two centuries. It has a strong tendency to sucker from the base. The plants are evergreen or deciduous leaves opposite or in whorls 5 or 6, short stemmed, oblong-lanceolate, 3 to 4 inch long, leathery. Showy flowers on the branch tips singly or as many as 5 in a cluster. They are 1 to 1 ¼ inch wide and characterized by the thick, tubular, red clayx having 5 to 8 fleshy, pointed sepals forming a vase from which emerge the 3 to 7 crinkled, red, white or variegated petals enclosing the numerous stamens. Nearly round, but crowned at the base by the prominent, clayx, the fruit, 2 ½ to 5 inch wide, has a tough, leathery skin or rind, basically yellow more or less overlaid with light or deep pink or rich red. All these characters make this shrub suitable for planting in the urban landscape especially in the arid and semi arid place with lesser maintenance.

Ghulam rasool, (2009) reported the status of pomegranate in Afghanistan, pomegranate is among the main exported fruits of Afghanistan that is exported to Pakistan, Russia and Arabic countries. Most of the pomegranate growing areas in Afghanistan are situated between 700-1500 m above sea level, and it can be grown in such ranges commercially. Forty eight (48) varieties of pomegranate available in Afghanistan, 20 varieties are sweet, 17 varieties medium sweet and 11 varieties sour. Among all of them seedless varieties very famous. The area of pomegranate orchards in Afghanistan is 2435 ha and the average yield is 9.5 t/ha. Generally Afghan pomegranate has 14 – 19% sugar content. Pomegranate is propagated by cuttings; the cuttings are taken at the end of winter and planted 20-30 days before the dormancy ends. Injuries to pomegranate due to low temperature are reported from Balkh province. Sun scald is one of the pomegranate problems in Afghanistan. Many growers prefer the multi trunk system wherever frost and sun scald are common features. Usually only one or two trunks are injured by these climatic factors. Single stem training is common in warm areas where there is no cold risk in winter. Fruit cracking is a main problem of pomegranate that is common in all pomegranate growing areas of Afghanistan.

Sezai Ercisli *et. al.,* (2009) revealed pomegranate cultivation in non-traditional northern Anatolia region of Turkey. Pomegranate cultivation is concentrated in subtropical regional in Turkey. However there are some special areas in temperate region where pomegranate grows well. The best examples of these areas are Coruh valley which is located in northern parts of Turkey. Coruh valley is accepted as one of the main plant biodiversity centers in the world by International Conservation Union. In this valley pomegranates are found as plant community with fig, walnut, cornelian cherry etc.

Finetto, (2009), says the pomegranate ranks 5th after grapes, almond, apricot and apple in Afghanistan. The area has decreased from 5667 ha (1996) to 2500 ha (2003). The average yields are in the range of about 8600 kg/ha in Dand district to more than 19000 kg/ha in Arghandab Mt, due to the poor production in some areas. Production in the Kandhar province is approximately 10,000 mt; and the main cultivar is Kandahari. Smaller volumes are produced elsewhere in Afghanistan. Elevation is the main restriction in production. With Bedana produced in Nangarhar at or above 1000 m elevation. Kandahari is produced at elevations from 550 m to below 1000 m, upto now no collection of local varieties has been carried out by the fruit tree projects funded by international organization. To meet the high demand of the world market, Afghanistan must increase production levels and improve productivity through better husbandry of pomegranate orchards. In Afghanistan few are common orchards, commonly the farmers grow this fruit species in small plots with intercropping, due to very exiguous dimension of farms (average0.8 ha per household), dire needs of staple food for their families on unstable political situation. Inspite of the good local varieties few cultural practices such as irrigation, fertilization, pruning, etc., are applied often in unsteady way, to the health benefits of the fruit (antioxidant) several traders in Afghanistan have started to receive orders for pomegranate from outside their traditional markets in the sub-continent. So, locally the market for pomegranate does not compare with other more common fruits because the demand is growing rapidly. Pomegranate commercialization has a typical supply chain that involves following major players; Grower, pre harvest contractor, Pakistani importer/Financier, packaging company. From harvest to market, wooden baskets (typically 3 sizes, but without uniform size/weight) made of flexible pomegranate tree branches (1st and 2nd year shoots) are used. For export mostly wooden crates are used. Pomegranate is ideal fresh fruit for Afghanistan to export as they came in the tough outer rind and had good properties. So, pomegranate should be considered as one of the most promising crops for support but refrigeration, post-harvest handling and promotion should be applied to this crop.

Chapter 2
Climate

Pomegranate has a versatile adaptability to wide ranging climatic condition. It can grow in the plains as well as on the hills up to an elevation of about 1840 meters. This fruit tree grows in low temperature, the tree is deciduous but in tropical and sub-tropical conditions it is evergreen or partially deciduous. The tree requires hot and dry climate during the period of fruit development and ripening. The optimum temperature for fruit development is about 38 degree Celceous. The tree cannot produce sweet fruits unless the temperature is high for a sufficiently long period. The quality of fruit is adversely affected in humid climate.

Pomegranate is winter-hardy and drought-tolerant plant and can thrive under desert conditions. Although, it is highly drought-resistant, the plant bears well only under irrigation. Aridity and frequent anomalies of the climate cause leaf shedding and fruit cracking.

Intercropping: In pomegranate the intercrops suitable are papaya, sapota, mango, drumstick.

Chapter 3

Soil

The pomegranate can be grown on diverse soil types. The tree can be grown in soils which are considered unsuitable for most other fruit crops. It can tolerate soils which are limey and slightly alkaline. The deep loamy or alluvial soils are ideal for its cultivation though it can be grown in medium or light black soils with minimum of 60 cm depth.

Edward Raja (2006) reported important micronutrient disorder in pomegranate in India. In Karnataka it grows in coarse textured Alfisols moderately heavy Inceptisols (clay loam in texture) and Vertisols with a clay soil in Maharashtra. The performance best in Inceptisols, moderately in the Alfisols and Vertisols, since they are varied in nutrient, a survey was conducted.

Tabatabaei and Sarkhosh (2006) studied the salinity tolerance of ten commercial pomegranate cultivars in Iran, with water salinity as main plot and 3 levels 4, 7, 10 dS/m and ten pomegranate cultivar as sub-plot. Vegetative growth, alive cuttings percentage after two months and percentage of established cuttings at the end of the experiment the necrosis and chlorosis level were measured during the experiment. The trees were cut at the end of experiment from 10 cm above the soil surface level and then the fresh vegetative weight and root dry weight were measured. Irrigation water, drainage water, soil in plots, root, stem, and leaves were analysed for their Na^+, Cl^-, K^+ contents.

Voshike-e-Saravan was the most resistant to salinity among the cultivars and Malas-e-Yazdi and Tab-o-Larz cultivars were also notified as salinity resistant cultivars.

Chapter 4

Cultivars Grown in India

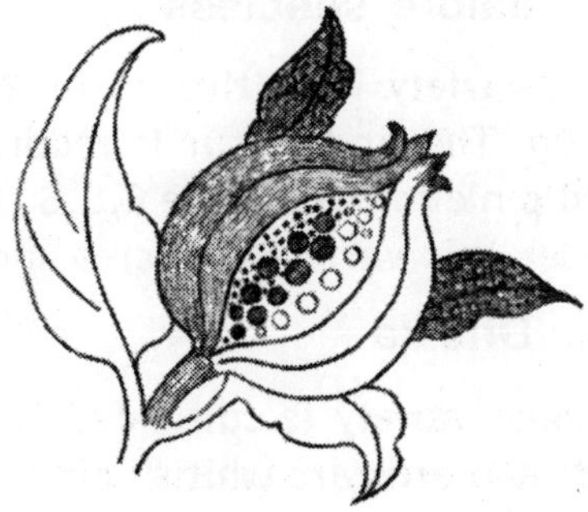

a. Varieties

i. Ganesh

Ganesh is a selection from open pollinated seedlings of Alandi. Earlier it was known as GBG-1. This variety has pinkish yellow to reddish yellow rind colour, having soft seeds and light pink arils fruits weighing between 225 to 250 gm, with T.S.S. 16 degree Brix with acidity of 0.3%. This cultivar has the highest cultivated area and very popular among the farmers and is also mainly exported. The arils of this cultivar attain pink color in winter months and are whitish in warmer months.

ii. Muscat

This variety used to be cultivated in Maharashtra and has fruit weight range of 300-350 g. The rind is yellowish pink; the grains are medium hard and whitish pink in colour. The T.S.S. is 16 ^{0} Brix with acidity of 0.5%.

At present this variety is not commercially cultivated and it not used for export.

iii. Bassein seedless

This variety is cultivated to some extent in Karnataka state. The fruit weight varies from 260 to 300 g. The outer rind colour is red; the seeds are soft having light pink coloured arils. The T.S.S. is 16 degree Brix with acidity of 0.37%. There is limited area under this cultivar.

iv. Jodhpur Red

This cultivar is from Rajasthan state. The fruits of this variety weigh between 180 to 220 g. The rind colour is yellowish red with whitish pink arils and hard seeds. The T.S.S. is 15 degree Brix with 0.6% acidity. This variety has no commercial importance.

v. Jalore seedless

This variety is cultivated in Rajasthan and has fruits weighing between 280 to 320g. The rind colour is reddish yellow to pinkish yellow. The seeds are very soft and pink coloured. The T.S.S. is 15 degree Brix with acidity of 0.42%. The area under this variety is increasing in Rajasthan.

vi. Dholka

Dholka variety is cultivated in Gujarat. It has red coloured fruits. The seeds are soft and arils are whitish pink. T.S.S. is 14 degree Brix with acidity of 0.41%. The fruits weigh between 280 to 310 g. This is grown in a limited area in Gujarat state.

b. Recent selections

i. GKVK-1 (Jyoti)

This variety is selected from Bassein seedless by UAS, Bangalore. The fruit is yellowish red in colour with soft seeds and deep red arils. The fruit weighs between 220-240 g and has T.S.S. of 15 degree Brix with acidity of 0.56%.

ii. IIHR selection

This variety is also a selection from Bassein seedless made by IIHR, Bangalore. The fruit of this variety is pinkish yellow and weighs about 180 to 120 g. It has light pink arils and soft seeds with T.S.S. of 13 degree Brix and acidity of 0.55%.

iii. G-137

This is a selection from open pollinated seedlings of Ganesh variety. This selection was made by MPKV, Rahuri. The fruit of this variety weighs between 250 to 280 g., is reddish yellow in colour. The seeds are soft and arils are light pink in color. The T.S.S. of this variety is 15 degree Brix with acidity of 0.36%. This is the only variety which has some acreage under cultivation in Maharashtra state.

iv. P-23

This is also a selection fro Muscat variety. The fruit of this variety is greenish yellow with pink tinge and weighs around 370 to 400 grams. The seeds are medium hard with light pink arils. The fruit has T.S.S. of 16 degree Brix and acidity of 0.42%.

v. P-26

This is also a selection from Muscat variety. The rind of the fruit is greenish yellow with pink blush. The fruit weighs between 360 to 400 grams. The seeds are soft with light pink coloured arils. The T.S.S. is 15 degree Brix with acidity of 0.41%.

vi. CO-I and Yercaud selection from Tamil Nadu state

CO-I is a selection made by TNAU, Coimbatore, having soft seeds. Similarly, Horticulture Research Station, Yercaud, selected a clone namely, Yercaud-I having medium sized fruits, easily peelable rind with soft seeds and attractive deep purple arils.

c. Hybrids

i. Mirdula

It is a selection from a cross between Ganesh and Gulsah Rose Pink varieties done by M.P.K.V., Rahuri. The fruit of this variety is red in colour and weighs between 230-270 g with blood red arils and soft seeds, T.S.S. of 17-18 degree Brix and acidity of 0.47%. This variety is in great demand from farmers.

ii. Ruby

Ruby is a variety released recently by IIHR, Bangalore. This variety is a multiple hybrid between Ganesh, Kabul, Yercaud and Gulsha Rose Pink varieties. The fruit weighs between 225 to 275 grams. The rind colour is red; seeds are soft with dark red coloured arils having T.S.S. of 17 degree Brix and acidity of 0.64%.

iii. Arakta

This variety is also a selection from open pollinated population of a cross between Ganesh and a Russian variety. The fruit of this variety is dark red in colour, soft seeds, dark red colored arils and high T.S.S. It is reported that it has acreage next to Ganesh and this is the most sought-after variety among farmers.

iv. Bhagwa

This variety has attractive glossy red thick skin with blood red and blood arils. It has also soft seeds and high T.S.S. This is also very popular among the farmers and is cultivated in large areas. It has also good shelf life and tolerant to fruit cracking. The Bhagawa variety of pomegranate presently under cultivation known by different names viz. Shendria, Ashtagandha, Kesar. The variety mature in 180-190 days (1) It fetches better prices which is 2-3 times higher than Ganesh, (2) Heavy demand for export (3) Suitable for long distance market (4) Better keeping quality (15-20 days at room temperature) (5) Moderately susceptible to black spots (6) Free from blackening of arils (7) No incidence of fruit cracking (8) No fruit drop (9) It gives high yield of 30-40 kg/tree (10) Tolerant to thrips, mites.

Suitability of Indian Cultivars for exports as fresh fruit

At present almost entire export is of cultivar Ganesh and Bhagawa grown in maximum acreage (90 %). The other varieties have small acreage namely, Arakta

(5 % of area), Mridula (2-3 % area), and G-137 (0.5 % area). The acceptability of varieties Arakta, Mridula, etc. is much in foreign markets because of very attractive fruit and aril colour.

World varieties of Pomegranate

Though we know a great deal about pomegranates and their origin there are a number of seedling varieties in the pomegranate areas that have no name selecting a variety with known qualities is the better choice. Most horticulturist divide pomegranate varieties into three categories Sweet, Sweet tart and sour, we prefer to include the seed hardiness in this and divide them into six groups.

1. Soft seeded sweet
2. Soft seed sweet tart
3. Early variety (mostly sweet)
4. Normal (harder) seeded sweet tart
5. Normal (harder) seeded sweet
6. Sour (Nearly always normal seeded)

The pomegranate varieties ruling in different countries are:

ARABIAN PENINSULA:

1. Romman chouall-Medium dark almost dark fruit, sweet tart flavour, grow in Iraq.
2. Mellassi- Large red fruit with soft seeds and sweet taste
3. Cherabani-Medium red fruit, sour taste, the syrup is used for preparing beverages and foods
4. Selimi-Large red fruit with sweet tart taste and small seed
5. Aswar (Black), Ahmar (Red) and Halwa also grown

AFGANISTAN:

1. Kandhar large white- Extra large yellowish sweet fruit, early ripening

BULGARIA:

1. Uzbekskii Saldikii (Uzbek sweet) sweet fruit\
2. Nikiskii Rannii-Sweet tart fruit early ripening and early to produce

CHINA:

1. Cin-Kwene-liu, Cin-pehin e-liu are some of the verities
2. Echen-large thin skinned (Paper shell) fruit, sweet taste

CYPRUS:

Zaitiki-large sweet fruit with soft seed

GREECE:

1. Patras acide-Extra large red fruit with very sour taste, good for syrup
2. Patras Douce: large red, sweet fruit productive

IRAN: Shahvar (Saveh) Red skin, very large fruit

2. Kandhar large red-Extra fruit with sweet flavour, mid season ripening. This is one of the best variety in the world of pomegranate
3. Kandahar large black-Dark purple (looks black) extra large fruit sweet tart fruit

ISRAEL:

1. Wonderful- The sweet tart variety from USA. High production with medium seeds
2. Asmar (Black)-Dark purple almost black fruit, white hard
3. Ras-el Bghil-Very large with normal seed
4. Red Lufani-Red large with normal seed good production
5. Malissi-Nearly seedless (called seedless) early ripening

ITALY:

Dolce, Alappi, Dolce Nostrale

MALTA:

1. Blanca: Medium whitish fruit seeds are undeveloped (nearly seedless) aril pink in colour productive
2. Dulce Colorada: Large red colour, red arils, soft seeded sweet, productive
3. Gorda de Jativa: Large red colour, red arils, soft seeded sweet, productive
4. Pignoncea: Large red fruit, small seed normal seeded sweet
5. Cagin: Large red fruit, very hard small seed sour, very good with sugar
6. Quiseppe: large, red, sweet fruit with high production

MEXICO:

Granada de china and Granada Agria are grown in Mexico

SPAIN:

1. Mollar-the best seedling pomegranate in Spain, sweet good red fruit, productive
2. Valencia: An early ripening sweet variety with moderate production

TUNISIA:

1. Gabsi-Large yellowish fruit, ripening with sweet taste
2. Tunsi-Medium red fruit, Dark red aril with sweet tart taste
3. Chelfi- Medium red, seedless variety
4. Djebeli- Very large red fruit, late ripening sweet variety with very small seed

TURKEY:

1. Hicaznar- A large good producing variety
2. AK-anar-Medium yellowish white, light seed and juice
3. Kizil anar- Large red fruit with sweet tart flavour, high production

UNITED STATES OF AMERICA:

i Wonderful

This variety originated in Florida, it has deep purple red fruit with medium thick and tough rind and arils are deep crimson in colour, juicy, with medium soft seeds. Wonderful cultivar has a T.S.S. of 17 ^{0}Brix or more and titrable acidity of 1.8%. This is a leading commercial variety. The fruit matures in late September and October. This variety ships well.

ii. Granada

This is a patented early maturing variety. It originated in California as a bud mutation of Wonderful. Fruit is darker red in colour with less acidity. The fruit ripens one month earlier than wonderful but smaller in size than Wonderful. Because of early maturity it commands a premium place in the market.

iii. Ruby Red

This variety has a limited commercial importance. The fruits are of the same size as Wonderful. It has crimson purple coloured fruits. It matures at the same time and reduces splitting of fruits. This variety does not store well as Wonderful.

iv. Early foot hill-**Large fruit similar to wonderful, ripens after Granada low acid and high sugars.**

YOGOSALAVIA:

Slatki Barshi Nar-Early ripening large, yellowish green fruit with sweet taste

Management Practices:

Rahim and Huq (2006) says pomegranate is a minor fruit in Bangladesh, recently its production and utility is increasing. Researchers at the fruit Tree Improvement Programme of Bangladesh Agricultural University have been working on pomegranate since 20 years ago. A number of varieties and production technologies have been developed which are used now for the expansion of cultivation of pomegranate in Bangladesh. Increased production of pomegranate fruits reduced the imports of this crop from abroad, thus, saving huge amount of foreign currency. This paper mainly deals with the present status and scope of pomegranate production in Bangladesh. Technologies such as tree management and post harvest technology and ICM that have been developed in Bangladesh will also be mentioned, possibly becoming useful to other countries that share similar climates.

Bar-yaakov, *et. al.,* (2006), says that, in an effort to select new cultivars that are most suitable for commercial usage, 44 pomegranate specimens were collected from different parts of Israel. These specimens are being grown in a single plot under identical conditions at Newe-Yaar Research Center. The pomegranates collected display prominent differences in their skin and aril colour, taste, fruit size, seed softness, date of ripening and growth habits. Out of the established population eight selections PG 100-1, PG 105-6, P, PG, 116-17, PG 128-29 and PG 130-131) were chosen and are being grown in commercial orchards. The early cultivars PG 128-29 and PG 130-31 ripen at the end of August and have a red skin and dark red sweet arils with soft seeds. PG. 118-19 and PG 116-17 ripen at the end of September and display a strong and appealing red skin colour and dark red sour arils. The established population of pomegranate trees at Newe-Yaar is used for breeding new pomegranate cultivars. We are now developing molecular techniques aimed to identify commercially important genes in order to select for promising cultivars as early as the seedling stage. The breeding program is focused particularly on selecting for early ripening and dark red skin and aril colour. Some promising crossbred progenies were selected and are being examined in a semi commercial plot.

Mohseni, (2006), summarized the situation of pomegranate production in Iran during the last 10 years, available commercial varieties whose fruit is used for export, as well as a survey of fruit production. He reports on major pomegranate exporting and importing countries, suggest economic market for the Iranian pomegranate production and discuss effective factors in marketing and exporting of fruit. Some of these factors include appropriate cultural practices, management of harvesting, and most importantly the development of commercial varieties based on a 5-year (2005-2009) Economic, Cultural and Social development plan, that are to add 6,297 ha of new commercial acreage and replant 15,655 ha of the old orchard acreage. The ultimate goal is to reach a total production of more than one million tons by 2009, increase fruit quality, and improve the processing and packing systems.

Ozguven, *et. al.,* (2006) says Turkiye is an area of origin of pomegranate. There are different types and forms of pomegranate have wide adaptation possibilities. Pomegranates are used for food as juice, canning, citric acid and animal feed, and for many industrial uses. Total pomegranate production of Turkiye reached approximately 80 thousand tons during 2003. Turkiye is an important pomegranate exporter country in the world. *Zizyphus spp., Prunus lauricerasus, Diospyrus lotus, Pinus pinea, Arbutus unedo* L, *Myrthus communis L, Pistacia entices, Juniperus oxicedrus L., Crataegus azarolus L., Cornus mas L, Morus spp, Cydonia oblonga,* and *Capparis spinosa* were also grown for centuries as a traditional fruits in Turkiye. They are spreaded and grown every areas such as in the coastal and the mountainous areas. This fruit species are grown as scattered trees or as road side hedges or as mixed with other fruit trees.

Kumar and Khosla, (2006), reported the performance of fourteen pomegranate cultivars viz., Achikdana, Anar shirin , Mohammed Ali, Chawla, Ganesh, G-137, Jalore seedless, Jodhpur Red, Kandhari Hansi, Khandhari kabuli, Nabha, P-75-K-3, P-75-K-5, P-23, P-26 was studied in the mid-hill zone of Himachal Pradesh. The average highest yields for five years were recorded with Chawla (23.2 kg/tree) followed by Nabha (21.9 kg/tree) and Kandhari kabuli (20.6 kg/tree) and the lowest with Jodhpur Red (13.2 kg/tree). The highest fruit weight was recorded with P-75-K-5 (508 g) and lowest with Nabha (329 g). The highest total soluble solids were recorded with Kandhari kabuli, which has the pinkish red fruit with blood red coloured grains and taste is sweet with acidic blend, is an excellent variety for cultivation of mid-hill zone of Himachal Pradesh, India.

Kumar, (2006), states that development of varieties such as Ganesh, Mridula (Arakta), Ruby, Bhagwa (Kesar) have revolutionized pomegranate cultivation in Karnataka state of India. These varieties have good export potential. Major break through in the cultivation of pomegranate such as resting period and pruning of plants, use of plant hormones to obtain better fruit size have contributed in increasing the production of quality pomegranate fruits to meet the preference of consumers in the domestic and international markets. Farmers of Bijapur, Bagalkot and Koppal districts have successfully exported pomegranates through their associations. Government intervention in providing technology and training has given a boost to the farmers to take up pomegranate cultivation for export purpose, also of late; problems due to diseases such as bacterial blight and wilt have had an adverse effect on the crop in the northern districts of the state. Government intervention in providing subsidies to the farmers, creating infrastructure facilities for production, post-harvest management and marketing and research support will go a long way in developing pomegranate crop in the dry areas and help the farmers in getting better returns.

Nasri, *et. al.,* (2006) reports about pomegranate geographical habitats and varieties and their ecological needs in Iran. Pomegranate is grown in Ivano-Turanian climatic region since ancient times. It is grown over an area of 338980 hectare with 381587 tones. The higher pomegranate production is recorded in Fars, Central (Arak), Isafahan, Yazd province. Pomegranate is considered as economic orchard crop due to the fact that its habits are usually located in central semi arid region of Iran. In Isafahan province for example the numbers of producers are 29309 hectare with production of 40968 tones which are mainly concentrated in Ardestan, Kashan, Shahreza region. Although the water deficit is a major agricultural limitation in these regions, the applications of traditional irrigation methods have resulted in the fine production quantity.

Ozezen (2006) reports about 'pomegranate' in Turkish culture its trees, its flowers, its fruits. A national point of view life style and perception is directly related to objects in the place don't exist cognition and conceptualization are not impossible what are we linking with which object and elements are we are forming between comparison and opposition which are two dynamic process for cognition. Then it is necessary that we must establish those objects with their functions and accordingly meanings. Because the names of mountains, rivers, cities, animals, fruits, vegetables, trees in any language mythical narration in connection with those names, the analogies to be made via those name, institualize expressions are positions of mirror of those society which use that language, in this context fruits and trees got a special important in Turkish culture, viz., mulberry, plum, apple, quince, cherry, grape, peach, cypress and willow etc. Pomegranate cultivation for export purpose, also oflate, problems due to diseases such as bacterial blight and wilt have had an adverse effect on the crop in the northern districts of the state. Intervention is providing subsidies to the farmers, creating infrastructure facilities for production, post-harvest management and marketing and research support will go a long way in developing pomegranate crop in the dry areas and help the farmers in getting better returns.

Chapter 5

Biotechnology and Plant Propagation

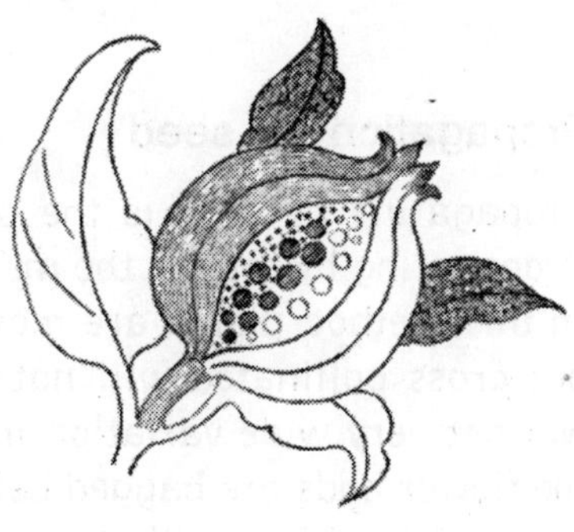

The use of biological system for the fulfillment of human needs perhaps started in 6000 BC when Sumerins and Babylonian fermented beer. Beginning with fermentation use of biological process, then experienced many changes over the centuries. However, latest revolution commence from 1970's and 1980's when a product of interaction between science of biology and technology came into wider existence this interaction got the name biotechnology.

Fermentation, Antibiotic production, baking the brewing are included in old biotechnology, whereas technique related with cell culture, fusion, bio-processing, genetic engineering were named as new biotechnology. Biotechnology in biological science represents such an interface of basic and applied science were gradual and septic transformation of science into technology witnessed.

Prokaryotes, Eukaryotic, Alga Glycophytes and Halphytes all are likely to contribute in commercially viable future proportion. Biotechnology is the application of biological organism, systems or process to manufacture and service industry comprises number of technology based upon increasing understanding of biology are the cellular and molecular level. The technique includes recombinant DNA preparation, tissue culture, protoplast fusion, protein engineering and immobilized enzyme, self catalysis, sensing with the aid of biological molecule. A fruit is a product of gynoecium with a accessory tissues involved of its life cycle the fruit passes through various development phases. One of this is fruit ripening. It commences when the growth of fruit is complete. The ripening of a fruit is associated with a series of perceptible changes involving the colour, the aroma, the consistency and the flavour. These contribute to making the fruit suitable for eating. Fruit ripening constitutes a natural irreversible process of aging correctly. It should, therefore, be referred to as fruit gain and fruit ripening. In many fruits, the aging process immediately after they are picked from the plant. For some fruits, picking is indeed an essential precondition for ripening. Others exhibit a marked acceleration of ripening compared to those left on the plant. Recent research shows that numerous biochemical processes run parallel in time during fruit ripening.

Propagation by seed

Propagating by seed is the oldest method of growing new pomegranate plants. Propagating by seed is the method used to hybridize new varieties of pomegranate. In this method plants are mostly true to the variety, if seed selections are proper but cross pollinated will not be true-to-type. Seed propagation in pomegranate will get very wide variation in the type of fruit produced. They are true-to-type if the flower buds are bagged before the blooms open. As pomegranate is self fruitful, some of the blooms that are bagged with a developed fruit and seed from such fruits should be true-to-type. In another method in many countries small fruits will be selected for propagation the reason behind this is that small fruit develop later in the season, where there were no pollinating agents around during the season, these small fruit is self pollinated and seeds should be true-to-variety. The seeds should be dried for two weeks for obtaining good germination.

Propagation by Grafting:

Pomegranate graft very easily and the best methods of grafting are cleft grafting, wedge grafting, whip and tongue grafting.

Bharati, (2006) reported the micro propagation of pomegranate conventionally through air layer or grafting. Micropropagation of pomegranate of soft seeded variety in large scale of recent introduction. Micropropagation technique have been standardized for large scale operation for the Indian varieties of pomegranates such as Bhagwa, Arakta, Rubi, Jothi plants produced in large scale and supplied to field planting. The adult tissues of apical and axillary buds were collected from field grown plants. The ex-plants are sterilized and in vitro culture were established in MS Media with different levels of plant growth regulators such as BAP, NAA, IAA pomegranate in vitro cultures were multiplied at 3 to 4 fold in every culture transfer and plants were multiplied into several thousands. The shoots of over 2 cm in height were rooted in vitro and transfer of hardening chamber. The tissue culture plants were hardened in the green house for 4 weeks and subsequently transferred to root trainer or poly bags. 30000 plants were produced and planted in the field and the performance of Micropropagated plants were assessed with that of air layered plants and found that tissue culture plants are superior than conventionally propagated plants.

Ramanjigowda, (2009) says Horticultural crops due to their high moisture content are inherently more liable to perish especially under tropical condition losses during post harvest operation due to improper storage and handling are high and can range from 10-40 per cent. With the advent of Bio-technological techniques resulting in use of molecular marker and transformation technique avenues have been opened for reducing post harvest losses and for improving quality of horticultural crops, Pomegranate fruit is neglected even though it is rich in vitamins, minerals and medicinal property, the Bio-technological intervention will help in improvement of the neglected fruit crops by adopting various techniques like plant tissue culture. Molecular marker and Genetic Transformation the Bio-technological development in pomegranate.

Ramchandra *et. al.,* (2009) said pomegranate is having major problems of wilt, two grafting methods wedge and tongue grafting and five dates (15th December, 30th December, 15th January, 30th January, 15th February) were evaluated during 2007-08 at National Research Centre on pomegranate, Solapur. One year old seedlings of the Phule, Arkata were used as root stock for the study. In general scion sprouting started between 8 and 12 days after grafting significantly higher scion sprouting was recorded with wedge grafting done in the last week of January after 15 (90%) and 21 (96.67%) days of grafting. Consequently maximum graft success (85%) was recorded after 90 days of grafting with wedge grafting done on 30th January. Plant height and scion length after 6 months of grafting was higher when wedge grafting was done between 30th December and 30th January. Grafted plants during this period had perfect union as indicated by normal diameter recorded at graft union, this reflect high scion and root stock compatibility. However performance of wedge grafted plants is to be evaluated under field conditions.

Zamani *et. al.,*(2009) observed conflicting reports on pomegranate (*Punica granatum L*). About the level of genetic diversity, genetic diversity is very low in fruits and others represent a high genetic diversity. Pomegranate wild accession were analysed by RAPD molecular marker. Completely different banding pattern were produced between the wild accession and cultivated ones, statistical analysis was carried out by NTSYS software and respective dendogram produced by Bootstrap software. Results represent high difference between the wild accession and other cultivated genotypes as its high similarity coefficient was 0.22 with others. In dendogram the accession separated from other genotypes in similarity distance of 0.15, this indicates that many cultivated accession of pomegranate which did not show a conspicuous difference in their genome level are progenies of common seed pattern or are derived by mutation during vegetative propagation. However non-cultivated type wild specimen may contain different genetic basis are worth while for higher consideration.

Irit Bar – Ya'akov *et. al.,* (2009) says pomegranate cultivars display a wide range of phenotypic difference such as growth habit, dwarfism, entrance to dormancy, chill requirements for dormancy release, content of secondary metabolites and fruit skin colour and incorporate parallel molecular analysis and establishment of cross bred population of pomegranate to produce colourful high quality fruits. In cross bred population using of parental lines combining appealing colours high quality fruits and other desirable horticultural traits. In pomegranate polyphenols and flavonoids particularly anthocyanins are the colouring molecules present in fruits and leaves. In order to study fruit colour in pomegranate the efforts were focused on isolation and characterization of the genes involved in the flavonoids anthocyanins biosynthetic pathway. Twelve genes were isolated by RT-PCR from pomegranate including structural and regulatory genes during fruit development. Most of the genes found are developmentally regulated during ripening at transcriptional level. Assessment of function of the individual genes is now studied under Arabidopsis model system. The molecular knowledge will enable to use identified genes responsible for colour variation as markers for selection in the cross populations.

Chaurasia *et. al.,* (2009) opined pomegranate (*Punica granatum L*) has been identified as one of the most important nutraceutical fruit with health enhancing properties. Molecular studies on the use of RAPD markers have enabled us to identify genetic variability among commercially cultivated India and exotic varieties obtained from major pomegranate growing areas such as Afghanistan, Iran and Turkey. In present study it was observed clearly distinguish different varieties using 24 random primers out of 52 primers that have been grouped together based on their geographical locations. The information generated by RAPD analysis can be used for studying genetic diversity and further improvements in the pomegranate.

Abdol Karim Zarei *et. al.,* (2009) reported Nine specific pomegranate SSR primers that had been studied previously were used for genetic analysis among 35 progenies derived from pollination of the famous Iranian pomegranate cultivar Malase – Torsue – Saveh. Most of these progenies were from open pollination, some are by selfing and a few from cross between Malase – Torshe – Saveh and soft seeded cultivar Bidaneh. Seven of these primers generated amplified fragments. Most of the produced bands were monomorphic between progenies and parents, only two polymorphic bands were detected between the samples by one of the primer pairs. Also two parents studied in this research were similar in SSR loci except for one. From these results it can be decided that these SSR loci are not enough for genetic discrimination between pomegranate genotype.

Swati Barche *et. al.,*(2009) observed that pomegranate is an ideal fruit for cultivation in all tropical to sub-tropical region due to its dietic, medicinal properties drought resistance wider adaptability, hardy nature low maintenance cost, high yield better keeping quality and push up to economy of farmers. Pomegranate propagation by seeds appears more hard but high degree of variability restricts its propagation. Nine treatment consisting of seven concentration IBA 2500 ppm, IBA 5000 ppm, NAA 2500 ppm, NAA 5000 ppm, IBA – 2500 ppm + NAA 2500 ppm, NAA 5000 ppm + IBA 5000 ppm and two levels of fungicide. Bavistin @ 0.5% and no fungicide were replicated thrice in completely randomized design.

Hard wood cutting were dipped for 30 seconds in prepared treatments. Cuttings were planted in polybags containing desired rooting media in partial shade. The results revealed that the highest number of shoots per cutting, length of shoot number of leaves per cutting. Survival percentage fresh weight of shoot, number of roots, length of roots, fresh and dry weight and dry matter percent of roots were recorded in IBA – 5000 ppm.

Jaganath *et. al.,*conducted investigation to know the effect of microbial inoculants on rooting of hard wood cuttings in pomegranate Cv. Bhagawa and Ganesh. Among different treatments *Pseudomonas florescence* + *Azotobactor chococcum* + *Trichoderm herzianum* viz., number of roots per cuttings (26.42), root length (19.80 cm) and root girth (3.01 cm) as well as sprout length (22.92 cm) sprought girth (0.75 cm), no. of leaves (46.77) and number of branches per cutting under low cost green house compared to open condition and other treatment combinations.

Jayanth *et. al.,*(2009) conducted study to assess the potentiality of microorganism in rooting of pomegranate stem cuttings, condition required for regeneration and their cost effectiveness in two pomegranate cultivars Ruby and Mrudula. The results revealed that the maximum percentage of rooting was obtained in Ruby (70) and Mrudula (63) in high cost polyhouse with IBA @ 2000 ppm, + Trichoderma @ 5 ml/kg pot mixture. Further there was insignificant increase in number of leaves per cutting at 30th day after planting (DAP) in IBA @ 2000 ppm + Trichoderma 5 ml/kg of potting mixture, while the number of branches (3.72) and plant height (18.98 cm) was significantly higher in IBA @ 2000 ppm + Azotobactor at 5 ml/kg of potting mixture on 45th DAP, but the diameter of shoot was significantly superior at 90th DAP in Azotobactor + Trichoderma each at 2.5 ml/kg of potting mixture. Cutting under high cost green house significantly increased the vegetative parameters like number of branches (3.59) on 30th DAP, length of shoot (3.72 cm) at 60th DAP, diameter of shoot (2.04 cm) at 90th DAP and plant height (20.21 cm) on 75th DAP.

Bhosale *et. al.,*(2009) studied response of different media and plant growth regulators on rooting and survival of air layers in pomegranate (*Punica granatum*) Cv. Sindhuri to find out best rooting media and PGR for better growth, development and survival of pomegranate air layers. The results indicate that sphagnum moss with IBA 5000 ppm recorded maximum days for initiation of root (17.33 days) while the maximum number of primary roots (14.00), secondary roots (40.00), root length (7.50 cm) fresh weight of root 0.73 gm and dry root weight (0.29 g) recorded with sphagnum moss with IBA 5000 ppm at 45 days after air layering. The maximum survival of air layer 73.33%, length of shoot (46.66 cm), number of leaves (52.33) leaf area (177.93 cm) and number of branches (9.0) with sphagnum moss with IBA 5000 ppm at 45 days after transplanting in nursery condition while the maximum survival of air layers (100%) length of shoot (46.66 cm) number of leaves (51.66), leaf area (171.06 cm) and number of branches (9.00) with sphagnum moss with IBA 5000 ppm at 45 days after transplanting in poly bag.

Vijay Bahadur *et. al.,*(2009) states that Effect of rooting media and indole Butyric acid on rooting and survival of cutting in pomegranate. Pomegranate is propagated through hard wood cuttings, in the propagation of pomegranate use of plant growth regulators play a major role to obtain better rooting and survival of cuttings. Rooting media also play an important role in root proliferation and development of nursery plants. The aim of the study to determine the most effective rooting medium and optimum concentration of IBA for rooting of pomegranate cuttings and their survival in the field. The maximum number of sprouts per cuttings (2.75), number of roots (60.50), root length (16.88 cm), root girth (2.60 cm) and survival of cutting (100%) in the open field were recorded with 750 ppm IBA treated cuttings grown on rooting media containing soil, sand, FYM in the proportion of 2:1:2 where as the maximum shoot length (24.75 cm), shoot girth 93.90 cm) and per cent of rooted cuttings (72.99) were noted with the treatment combination of 750 ppm IBA with rooting media, soil, sand and vermicompost in the ratio of 1:1:2. The minimum days for sprouting in cuttings (10.80) were noted with the treatment combination of 500 ppm IBA with rooting containing soil + sand + FYM 2:1:2. The maximum number of leaves per shoot (50.00) was recorded with 1000 ppm IBA and rooting media with soil + sand + FYM with a ratio of 2:1:2. The maximum benefit cost ratio (2.75) was obtained with 750 ppm IBA with rooting media soil, sand, and FYM in the proportion of 2:1:2.

Bikramjit Singh *et. al.,*(2009) elaborates about influence of planting time and IBA on rooting and growth of Pomegranate (*Punica granatum L*) cuttings Cv. Ganesh where uniform size cuttings of pencil thickness, 20 cm long with three to four buds were taken from healthy pomegranate trees in December, January and planted during December 15 and January 15. Growth regulator (IBA) was applied by adopting quick dip (for 5 seconds) and slow dip (for 24 hours) techniques. The respective cuttings were treated with varying concentrations of IBA 0, 50, 100, 200 ppm for slow dip and 1000, 1500, 2000 ppm for quick dip techniques and planted on different times which showed its effect on sprouting, rooting behaviour and plant growth. Treated cutting were synchronized with better root formation and shoot growth potentialities. Moreover IBA 100 ppm and 2000 ppm were found to be most efficacious in encouraging rooting and invigorating shoot growth. The cutting treated with IBA 100 ppm (slow dip) and 2000 ppm (quick dip) planted in the month of January exerted positive effect with regard to sprouting percentage, rooting percentage, number of roots per cutting, longest root, root weight, plant height, shoot girth. The cuttings planted during the month of December also resulted in good sprouting.

Aytekin and Caliskan (2009), Rooting of stem cuttings of five pomegranate types selected from Delibekerili, Turkey taken in different periods under mist propagation were investigated. The cuttings were prepared 20-25 cm in length from wood shoots and 1000 ppm IBA was used as a rooting hormone. In the study rooting ratio (%), root number, root length of the cutting were measured. Among the cuttings rooted the highest rooting percentage was observed in 31-N-01 type. There were higher roots in the 1000 ppm IBA treated cuttings than the control cuttings for the 31-N-1, 31-N-02, 31-N-03 type where the 31-N-04 and 31-N-05 type had lower roots in the 1000 ppm IBA treated cuttings than the control cuttings. 31-N-1 type when compared to other gave more favourable results in terms of root length. The 1000 ppm IBA treated affected positively rooting percentage of cuttings and the other characteristics although it was not to the satisfactory level. It was concluded that increasing dose of IBA could be useful.

Genetics resource management and crop improvement

Pomegranate in its diploid set has 16, 16-18, 18, 18-19, chromosomes. Darlington believed that in angiosperm plants the haploid number of chromosomes is 7. G. V. Dergach determined that in socotria $2n = 14$ which is why the haploid number of chromosomes $n = 7$ in Punicaceae can be considered a more primitive characteristics from evolutionary stand point while $n = 8$ in punica as a factor of advancement. Pomegranate chromosomes are very small (1-2 μm) and differ significantly in size.

Mutations: Pomegranate is characterized by insignificant mutation ability. Spontaneous somatic mutations in pomegranate are expressed in change of fruit color, their quality, ripening terms crops capacity and plant development.

Wonderful and papershell were used as donor for soft seededness. Improvement of pomegranate can be done by three methods (1) collection of superior germplasm from indigenous and exotic sources (ii) improvement by selection (iii) controlled hybridization. Collection of superior germplasm was initiated in Maharashtra State in India in 1944 but collection of exotic cultivar was done from Iraq, Afghanistan, Iran, Russia and other countries.

Purohit (1983) reported that attempts to introduce superior deciduous cultivars in the tropics failed because of development of disorders caused by lack of temperature below 7°C in winter and absence of dormancy. The deciduous cultivars hardly produce 5-10 flowers per plant per year, in tropics and out of this 75% flowers male and intermediate and hence do not set fruits. Therefore improvement of evergreen pomegranate suitable for tropical regions was initiated in 1980 in Bengaluru.

Improvement by selection, Hybridization and cross pollination are the two important factors which brings about the variation in the seedlings and this naturally induced variation is used for the selection of good types. Selection of seedling from Alandi was done by Dr. G. S. Cheema and gave rise to GBG-1 (Ganesh) in respect of prolific yield, medium size fruit with soft seeds, better pulp and sweet juice with an extremely agreeable flavour is not comparable to imported fruits of exotic cultivars particularly in regard to size and quality. Sawat (1973) made an attempt to identify desirable types of clonal selection from Ganesh pomegranate and reported clones G-137, G-133, G-107, G-132 as superior to Ganesh. The success in breeding depends on extent of genetic variation present in various biometric characters in plant population. The assessment of variability is the first step in breeding programme. Hybridization in pomegranate with introduction of Russian cultivars was collected in 1973, but didn't suits to our environment, but they had red colour of rind and aril. Ganesh was crossed with four Russian cultivars Shirin Anar, Gulsha Rose pink, Gulsha Red, Gulsha.

Indian Institute of Horticultural Research, Bengaluru developed a multiple hybrid Ruby, which develops red aril colour in summer too. Of late Bhagwa which is a selection from farmers field in Maharashtra, occupied large area in India, with synonym Ashtagandha, Sindur, Shendria and Kesar in different regions, but all the varieties developed in India are susceptible to Bacterial blight (*Xanthomonas axonopodis* pv. *Punicae*), no variety in India is resistant to bacterial blight, which has become a major problem in cultivation of pomegranate in India. Developing varieties / hybrids resistant to bacterial blight is the only solution to overcome this problem and the work is under progress in different research stations in India.

Malliaradhaya *et. al.,*(2006) studied low genetic diversity and lack of genetic differentiation in punica as evidenced by amplified fragment length polymorphisms. An amplified fragment length polymorphism (AFLP) based analysis of genetic structure and differentiation using 65 accessions of pomegranate representing the central and west Asia region. Russia, USA revealed that low genetic variation and lack of population structure. Accessions were evaluated across 356 polymorphic fragments generated by 15 Eco RI and Mse I primer combination. Three some what distinct clusters along with many minor groups were evident from a neighbouring joining cluster analysis with marginal differentiation among the clusters. There were many significant association from Turkmenistan such as chico sour, Phoenicia from the US with the cultivar Azadi, Haku Butan from Japan with a cultivar seidi and the leading US cultivar wonderful with two juice type cultivars Kolobok, Myatadzhy partitioning of molecular variation suggested that nearly 65% of the variation resides within the group and the remaining variation was due to differentiation within (23%) and among groups (12%). Surprisingly *Punica granatum* exhibited a wide range of adaptability across the tropics and sub-tropics; the analysis suggests a narrow variation and lack significant genetic divergence within the genus. Potential implications of the results due to future prospects of genetic improvements and range of adaptation.

Singh *et. al.,* (2006) studied conservation, characterization and evaluation of pomegranate germplasm under and ecosystem of India. Versatile adaptability, hardy nature, low orchard management cost, high yield and excellent keeping quality have made pomegranate fruit more popular in arid and semi arid regions, 30 genotypes from different parts of the world were evaluated, propagated by cutting and planted at a spacing of 5x5 meter. Wide range of variability was noticed for different vegetative, flowering, fruiting characteristics. Sixteen genotypes behaved evergreen and rest 14 as deciduous. Total soluble solids contents was highest in cv A.K.Anar (21%) followed by Speen danedar (20.5%). Maximum titrable acidity was observed in Tujetis EC – 104347 (3.41%), and Dorsetta Malus (3.70%), cultivar Mirdula yielded maximum juice recovery (52.6%), fruit cracking was least in Kazaki anar (10.0%), Patna-5 (11.0%), Kabul (10.1%), Gul-e-Shah (12.0%) and in Jalore seedless (15%). The result of organoleptic testing of juice for flavour was in range of very sour to sweet. Juice color varied from light pink to deep red. The evergreen types are suitable to hot climate.

Shephavand *et. al.,*(2006) studied evaluation of the morphological and fruit characters, cold hardiness of ten pomegranate accession cultivated in Karaj Iran. Pomegranate is mainly grown in areas where winters are cool and summers are hot. Ten pomegranate accession were studied to detect similarities among genotypes and assess their performance and adaptation under Karaj climate. Many different qualitative and quantitative characteristics were measured during 4 years. Factors and clusters analysis showed a considerable phenotypic diversity in local pomegranate accession. Analysis of variance showed that there was a significant difference in cold hardiness between different genotypes and some genotypes were resistant than others.

Manmohan Sharma *et. al.,*(2006) studied genetic variability in Wild Pomegranate in Rajouri and Pooch district of Jammu and Kashmir.

Wild pomegranate (*Punica protopunica* Dalf.f) growing in the intermediate zone of Jammu and Kashmir at lower and upper hills of Western Himalaya is an erect deciduous shrub with spreading growth habit, bearing attractive red flowers. Fruits are globular crowned by a persistent calyx and have a hard yellowish green rind with a red tinge. The seeds are angular with fleshy aril which constitutes the edible portion. A random survey was conducted for the extent of variability with respect to various fruit tract and yield. Per cent juice content of fruit yield per tree was above average when fruit weight was moderate. Rind thickness, fruit size and core diameter had a high level of variability in the germplasm.

Varasteh *et. al.,*(2006) studied the evaluation of the most important fruit characteristics of some commercial pomegranate cultivars of Iran. Five important cultivars of pomegranate characteristics in cv. Malas-e-Saveh, Rabab-e-Neyriz, Malas-e-Yezdi, Sishe-kape-Ferdows, Nadery-e-Budrood were selected and labeled for further measurements in laboratory. Fruit diameter, length, weight, volume, skin thickness, juice yield, aril yield, weight of 100 aril weight, total soluble content, pH and titrable acidity were recorded in 2005. Recorded data may be used to develop an index for cultivars selection in pomegranate production and breeding in future.

Nobigol and Talaei, (2006), says Iran is pomegranate centre of diversity and it has valuable genetical resources, which can be used in breeding. The description of pomegranates germplasm has been done due to its morphological and pomological criteria, and limited; studies have been done on it. Today the modern breeding systems due to which plant breeding has proceeded faster and more accurately, are discussed. It should be considered that pomegranate growth in Iran faces many problems and recent improvement in genetic and biotechnology can be an effective mechanism for solving these problems. More research should be done on this field and develop new methods for cultivars identification and genetic resources management is necessary.

Sharifani, *et. al.,*(2006) studied the diversity in four populations of wild pomegranate in four regions of Golestan province. From each area, four trees were selected; eight variable including leaf length, leaf width, flower length, style length, flower diameter, flower peduncle, leaf petiole length and internode length were selected. Data of ten replications for each variable were recorded. All measurements were achieved by using calipers. The data analyzed in a complete randomized block design using a SAS software and Duncan test ($P<0.01$). Results indicated that flower length, flower peduncle and internodes are significant characters and internodes are significant characters for evaluation of the populations. The characteristics of wild samples from the two areas were similar. The results showed the four populations under study were divided into 3 classes. Further study is underway, using RAPD technique to test the recent results.

Muradoglu, *et. al.,*(2006), says pomegranate bushes are wild grown in cukurca district (Hakkari) located in eastern Turkey. In this study, desirable pomological traits of some pomegranate genotypes from Cukurca district were described. Genotypes had a range of 131.337 g for fruit weight, 60-81 mm for fruit height, 30-88 mm for fruit width, 18.83 g for 100 stone weight, 1.7-4.0 mm for fruit shell thickness, 8.8 – 16.7% for soluble solids content, 0.2 – 1.5% for acidity and 1.1 – 2.6 cm for calyx length. On the other hand, other important fruit characteristics were also identified.

Gentile *et. al.,*(2006), conducted a study with the aim of containing the genetic erosion of *Punica granatum L.* and of characterizing its genetic diversity in Sicily, eight accessions individuated in rural areas of eastern part of the Island have cultivated in an experimental field of Catania University together with the most important cultivar and with four Spanish varieties. Bioagronomic features of the plants and pomological and analytical characteristics of the fruits of these 13 accessions have been evaluated for four years. The results indicated that among the local Sicilian accessions exists a relevant variability degree concerning the fruit characteristics; (i.e., seed hardness, fruit colour, ripening period) but, apart from the accession named primosole, which showed good bio-agronomical performance, none of them appears to possess overall satisfactory features in comparison with the Spanish varieties. These on the contrary exhibited good performance and a less variability degree on the whole, the intraspecific genetic diversity already individuated, together with other uncharacterized accessions, represent an important source of germplasm for new breeding programs for the utilization and products of this species.

Lamalfa, *et. al.,*(2006) carried out a selection work in Sicily in order to characterize the local germplasm and to find new genotypes with good behaviour also in comparison with varieties coming from abroad. Five accessions individuated in rural areas of the east part of the Island have been compared with the most important Italian cultivar (Dente di Cavallo) and with four Spanish cultivars (Pinonenca, Valenciana, PiNon tierno, Moller de Elche). An accession of local germplasm named 'primosole' showed good bioagronomical performance and is proposed for further diffusion among the local Sicilian accessions, a relevant variability degree was revealed concerning the fruit characteristics some of these genotypes showed interesting traits i.e., seed hardness, fruit colour, ripening period) but, apart from primosole none of them appears to possess overall satisfactory features in comparison with the Spanish cultivars. These on the contrary exhibited good performance and a minor variability degree on the whole, the intraspecific genetic diversity already individuated, together with other uncharacterized accessions, represent an important source of germplasm for new breeding programmes for the utilization and products of this species.

Islam, *et. al.,*(2006), carried out a study during 2005 in Trabzon, Giresun and Ordu. The aim of the study was to put forward to pomological and morphological characteristics of pomegranates grown in Black sea region. The results shown, fruit weight is between 123 g and 295 g, skin thickness is between 2.95 mm and 6.42 mm, seed number per fruit is 103-667, total soluble solute is between 12%, fruit size is medium or large, flavour is sweet or sweet sour and skin colour is generally reddish yellow. The types are mainly used in fresh consumption.

Ozguven, *et. al.,*(2006), conducted a study with a aim to determine the adaptation of 10 pomegranate cultivar to the North Cyprus ecological conditions. They carried out with 10 pomegranate varieties (33N 11 Asmar, 33N 12 EksiNar, 33 N 16 Silifke Asisi, 07N 14 Mayhos IV, 07N04 Mayhos II, 07N 03 Yufka Kabuk, 07N 08 Hicaznar, Izmir 1265, Izmir 1264, Izmir 1) in the Turkmenkoy Research Station in the North Cyprus by Department of Horticulture, Faculty of Agriculture, University of Cukurova between 1999-2003. In the study the yield, phenological and pomological characteristics of pomegranate cultivars were determined. Several pomegranate cultivars were found to be suitable for the North Cyprus and Hicaznar was determined as most yielded cultivar.

Mehranna Koohi Dekhkordi *et. al.,*(2006) studied the characterization of microsatellites and development of SSR markers in pomegranate. Pomegranate a excellent tree for arid and semi arid region of Iran. High diversity was observed in Iran. Pomegranates are known by local names and distinguished based on morphologic characters. Because of codominancy, mulliallelism characteristic and even distribution in genome, SSR markers have been used for classification and genotyping of different germplasm. Since no SSR primers have been available for pomegranate its microsatellite were studied and isolated. Edwards method used for the study which is based on partial genomic library construction and enrichment in order to capture digested fragments containing microsatellite. Primer pairs for only 12 microsatellite could be designed, since the flanking sequence of the others were either too short or were otherwise not suitable for designing the microsatellite specific primers. Microsatellite loci are the expected size and polymorphism were successfully amplified from 4 of these 12 primer pairs using 29 pomegranate cultivars.

Ataseven Isik (2006) studied the similarities between pomegranate parents and crosses regarding some fruit characters. After self pollination and crossing of two sweet and one sour sweet standard pomegranate varieties, 2487 types obtained from 9 combinations were compared with their parents regarding taste, rind colour, aril colour and seed harness properties and similarity percentage of each combination were determined. The rind and aril colour were medium FellahYemez X Ernar 87.63% and 93.58% of Ernar X FellahYemez combination observed to be sweet. On one hand the higher percentage (37.11%) of the types in FellahYemez X Ernar combination were soft seeded where as the higher percentage (44.92%) of the types in Ernar X FellahYemez combination were hard seeded. 39.70% and 40.66% of Hicaznar X ERnar and Ernar X Hicanznar combination were sour and also 57.33% and 52.33% of Hicaznar X Ernar and Ernar X Hicaznar were hard seeded respectively. While 38.50% of the types obtained from Hicaznar X FellahYemez combination were soft seeded, 44.64% of the types obtained from FellahYemez X Hicaznar con bination were found to be hard seeded.

Sarkhosh (2006) studied the relationship among quantitative and qualitative characteristics of fruit components of pomegranate genotypes.

Sarkhosh *et. al.,*(2006) studied the relationship among quantitative and qualitative characteristics of fruit components of pomegranate genotypes seed and fruit were studied to evaluate 24 pomegranate genotypes for total soluble solids, titrable acidity, Anthocyanin, pH, 100 aril fresh and dry weight, 100 seed fresh and dry weight, length and diameter of aril and seed, peel thickness were evaluated. Analysis of variance revealed that all the investigated characters between examined genotypes were significant showing variability on characters between genotypes. Results of simple correlation analysis showed the existence of significant positive and negative correlation among characters (e.g., significant positive correlation between aril 100 weight, and seed length, significant positive correlation between peel thickness and diameter of fruit neck. Factor analysis performed for determining the main characters showed that analysis using the seven factors that explained 89% of total variance divided genotypes into 5 sub clusters at distance of 9. Further more by using the three first manfactors, genotypes position were determined in 3 – plot analysis. This study showed that multivariate analysis is a useful method for discrimination of pomegranate genotypes.

Hasnaoui *et. al.,*(2006) reported about construction of an enriched genomic library of pomegranate and design of SSR primers. Microsatellite or SSR markers were demonstrated to be very useful for the detection of variability within genera as well as species because of their high level of Polymorphism, reproductivity, co-dominance and abundance. As pomegranate genome is poorly studied a new genomic library made of more than 480 colonies which were supposed to contain microsatellite sequence was constructed. Among these 172 have been transformed by an insert containing microsatellite pattern. Then inserts were extracted and sequences were performed. Sequence microsatellite receptive core regions as well as the flanking region were determined. The primers design was made using the primer 3 software. It was possible to perform more than 50 potential SSR pairs of primers and 25 pairs were tested on 4 pomegranate cultivars. Finally 22 pairs were modified to be used in Mega BACE for the screening of Tunisian pomegranate genotypes.

Hasnaoui *et. al.,*(2006) studied molecular polymorphism in Tunisian pomegranate germplasm as revealed by Rapid finger prints. The genetic diversity and polygenetic relationship among Tunisian pomegranate cultivars have been studied using universal primers the Random amplified polymorphic DNA (RAPD) was designed to generate banding profiles from a set of twelve local cultivars. Data was completed with appropriate programme to construct a dendogram illustrating the relationship among the germsplasm. Results proved the efficiency of this method to examine DNA polymorphism in pomegranate. Thus tested primers are unauthorized by collective resolving power (Rp) of 12.83. In addition cluster analysis exhibited parsimonious tree branching made independently from the geographic origin of the cultivar. The RAPD constitute an appropriate procedure to assess the genetic diversity and polygenetic relationship.

Jbir *et. al.,*(2006) studied molecular polymorphism in Tunisian pomegranate cultivar as revealed by AFLP. In Tunisia pomegranate germplasm is characterized by large diversity since more than 60 cultivars were identified; and little is known about genetic diversity in the Tunisian germplasm. The use of morphometric and agronomic parameters mainly related with fruit traits and proved their suitability in the cultivars discrimination. Molecular studies were undertaken using Amplified Fragment length polymorphism (AFLP) markers. The development of different primers combinations in selective amplifications has permitted detection of a large number of molecular markets suitable for genetic diversity surveying. Data showed high polymorphism at DNA level. This study portrays the development of AFLP as a powerful procedure reliable in the establishment of Tunisian pomegranate cultivars finger printing. Results are discussed in relation with the establishment of selection program and rational management of the local germplasm conservation.

Yilmaz., *et. al.,*(2006), worked on the molecular characterization of genotypes in pomegranate revealed that, molecular markers can detect differences among varieties directly at the DNA level. RFCD, RAPD, ISSR, AFCD and SSR markers are using for cultivar identification. In this Random Amplied Polymorphic DNA analysis was performed on some pomegranate genotypes which were selected from different regions of Turkey. One hundred and twenty primers were applied to morphologically different four genotypes coming from different origin for screening and only five primers produced polymorphic bands and they were selected to use for RAPD reactions with all genotypes. Genetic similarity values were calculated and unweighted pair group method analysis) cluster analysis was performed to generate dendograms. RAPD analysis was not able to distinguish some variety/ type. This study shown that pomegranate genotypes have very narrow genetic basis according to RAPD analysis and studied genotypes. This study will continue with the other marker systems.

Zamani *et. al.,*(2006) studied the RAPD Markers for assessment of genetic variation between pomegranate genotypes in Iran. Clear identification of pomegranate is a difficult task for different purpose especially for breeding purpose. RAPD markers were used for studying 24 genotypes, some of them having similar method and a number of 113 decamer primers were tested on them, among which 27 showed good amplification and produced 158 polymorphic bands. Cluster analysis using Jaccarda similarity coefficient and UPGMA method divided these genotypes into sub cluster at a distance of 0.64 on the dendrogram. Genotypes with similar names did not perform a high similarity and scattered into different sub-clusters. Geographic origin of genotypes affected the clustering to a high extent, while characteristics such as fruit colour or taste did not show prominent effect. The observed variation was approving the richness of pomegranate genetic resource in Iran. Also RAPD markers performed as a technique for pomegranate genetic studies.

Jalikop *et. al.,*(2009), Pomegranate cultivation is one of the most attractive farming enterprises in the Indian arid tropics. Aril browning or internal break down of aril in pomegranate fruit has become a serious concern to consumers, growers and researchers in the recent time. In order to understand the genotypic variation for aril browning and its association with other fruit traits 159 progenies obtained by selfing two pomegranate multiple hybrids viz (Ganesh & Kabul) x Yeracud x {(Ganesh x Gulsha Rose pink) – F2} and {(Yeracud x Jyothi) x Ganesh x Guylsha Rose pink F2)} were studied. Because of diverse genetic base of parents and heterozygous nature of the crop, a wide combinations were produced which were scored for aril browning, fruit skin colour, total soluble solids and seed mellowness. Results of spearmans correlations analysis revealed that aril browning is inversely related with skin r = -0.15 and aril (r = - 0.41) colours. A statistical model constructed to study the reasons for the observed variation in aril browning showed that about 82.9% of it is accounted collectively by skin colour, aril colour, Tss and seed mellowness. Further a refined model represented by Y (aril browning severity) = 0.78 – 0.52 x 1 (aril colour) + 0.23 x2 (Tss) would contribute for 3.5% of the observed variability in aril browning with least error in prediction (mean square error being 0.31). Studies showed that a unit increase in intensity of aril colour amount to proportionate decrease in severity of aril browning to the tune of 0.52 units; however for every unit increase in Tss there will be increase of 0.23 units in severity of aril browning. Thus the increase in intensity of skin colour and aril colour there will be reduction in severity of aril browning while with raise in Tss aril browning incidence will be more, an association not favourable in selection of desirable genotypes. Breeder should therefore strike a balance in aril colour and Tss level while locating recombinants free from aril browning.

Ozguven *et. al.,*(2009) conducted a study on fruit traits of some pomegranate cultivars in adana ecological conditions with the aim to determine the adaptation of 21 pomegranate cultivars Yzmir-1, Yzmir-2, Yzmir-8, Yzmir-10, Yzmir-12, Yzmir-15, Yzmir-16, Yzmir-23, Yzmir-26, Yzmir-29, Yzmir-1261, Yzmir-1264, Yzmir-1265, Yzmir-1267, Yzmir-1445, Yzmir-1453, Yzmir-1465, Yzmir-1479, Yzmir-1483, Yzmir-1499, Yzmir-1513, cvs, fruit weight, fruit length, fruit diameter aril rate, juice rate, peel thickness, Tss, acidity, skin colour, taste, seed hardness were determined at fruits in pomological analysis, several pomegranate cultivars were found to be suitable for the ecological conditions of the adana and fruit characteristic varied in cultivars.

Rakesh Bhargawa *et. al.,*(2009) says flavonoids as an acid in cultivar identification and assessing phytogenetic Kinship among some cultivars of pomegranate. Secondary metabolites and flavonoids gained reputation as being excellent phyto chemicals markers for cultivar identification. These markers come hardy at places where the morphological parameters become limiting and hence recourse to such phytochemical markers are taken to delimit the taxon. Pomegranate is represented by a large number of cultivars which have been identified / developed for specific purpose. The genetic variability once generated has been fixed on account of vegetative propagation which is commonly practiced in taxon. Varietal identification of these cultivars is based mostly on fruit character. These at times become highly overlapping and hence pose problem in varietal delimitation and assessing the phylogenetic Kinship between the cultivars which is essential for pre-breeding programme. Accordingly the data generated revealed that flavonoid spots or spot combination is specific for each cultivar which can be used for not only for varietal identification but also for assessing the phylogenetic Kinship among pomegranate cultivars. Thus pomegranate cultivar Kabul could be identified by spot no.9 where as cultivars which could be identified by unique combination were Mirdula (18, 20, 24), Jodhpur red (4, 14 and 18), P-23 (4, 14 and 15) and Ganesh (1, 3, 18 & 19).

Jalikop and Kumar, (2009) says double flower pomegranate originated from a hard seeded acidic pomegranate by spontaneous dominant gene 'DF' mutation. An ornamental variant of pomegranate known as Double Flower produces attractive large rose like flower due to phenomenon of petaloidy where in the stamens are modified to petals. Trees with such flowers are incapable of setting fruits on their own. When the double flowers pollinated from pollens of cultivated pomegranate they produce fruits suggesting that absence of fruitset in Double flowers is due to functional male sterility, by studying the genesis and genetic bases of petaloid nature by crossing normal or single flowered genotype (which breed true to type for single flower) viz., Ganesh, Ruby Red, Kabul Yellow (Sweet types) Nana, Daru and Ganesh & Kabul Yellow (Sour type) with Double flower variety. A total of 520 Double flowers were pollinated with the pollen of above varieties and hybrids. Fruit set (14.87%) seed germination (5.23%) and establishment of F_1 hybrids (36.48%) were low. The F_0 fruits (produced from the double flowers) were morphologically different with much of the ovary exposed and they contained pink, highly acidic, hard seeded aril. Out of 112 F1 hybrids, 109 flowered and among them 45 were single flowered and 64 Double flowered which was a good fit for 1:1 ratio (c^2 = 3.31, p = 0.05 – 0.10) suggesting that a spontaneous dominant gene mutation (Df) in a hard seeded acidic pomegranate caused modification of stamens in to petals. Further in Kabul yellow x (Double flower x Nana) among 32 individuals 16 were single flowered and 16 were double flowered which fitted perfectly in to genetic ratio of 1:1 confirming the monogenic inheritance and dominant nature of double flower. In pomegranate acidity is monogenetically controlled with higher fruit acidity (sourness, SS ss) being dominant or low acidity (Sweetness, ss). All hybrids between the double flower and Ganesh should have been sour fruited in Double flower variety is homozygous for fruit acidity or if heterozygous half the progeny should have been sweet and half should have been sour. But in F1 families of Double flower x Ganesh (Dfdf x dfdfss) ten single flowered progenies produced fruits all of which are sweet. Hence the genetic constitution of Double flowers DfdfSs and this genetic integrity is maintained because of clonal propagation.

Joshi *et. al.,*(2009) reported on genetic diversity in pomegranate (*Punica granatum L*) var. Bhagawa. The divergence analysis among sixteen Bhagawa types of pomegranate collected from different pomegranate growing pockets of Maharashtra was worked out using Mahalanobic D^2 analysis, based on D^2 values five clusters were formed among the sixteen types. Out of this the cluster 'A' was the largest having six genotypes followed by the cluster 'B' and cluster 'C' four genotype and cluster 'D' and cluster 'E' having one genotype each. This indicates the wide divergence among the genotypes. The D^2 value ranged from 193.69 to 6484.16. The minimum D^2 value of 193.69 was noted between the genotype Sel.15 and Sel.9 while it was observed maximum 6484.16 in genotype Sel.3 and Sel.12. The inter and intra cluster D^2 values indicated the maximum inter cluster C and D (8.32) followed by cluster C and E (8.17) and observed lowest between cluster B and D (5.53). The cluster mean for average plant height was maximum in cluster B (2.66) and was minimum in cluster C (2.26). The cluster mean for days required for fruit set to maturity was maximum in cluster B, D and E (183 each) and registered lowest in cluster A (171.50). The cluster mean observed minimum in cluster B (42.25) in respect of fruits per tree. The value of cluster mean in respect of fruit yield (kg/tree) was maximum in cluster D (12.62) followed by cluster E (12.48) and lowest in cluster B (11.65) in respect of average fruit weight, the cluster mean was maximum in cluster E (284.82) while it was minimum in cluster C (265.25) where as cluster value in respect of rind thickness was maximum in cluster D and cluster E (0.40) and it was found minimum in cluster A (0.33). Cluster D showed maximum mean value (49.96) in respect of juice content and 15.72 in respect of Tss% of fruit. The value of cluster mean in respect of acidity was observed maximum in cluster E (0.41) and minimum B and C (0.38). The same trend was observed in respect of disease (wilt) and pest (thrips). Hence the isolated genotypes are grouped in same cluster i.e., cluster C.

Yilmaz *et. al.,*(2009) said pomegranate (*Punica granatum* L) is a native and traditional species in Turkey, it is spread and grown in every area such as in coastal and mountainous areas as much as 1000 meter high altitude. Mediterranean, Aegean area, South eastern regions in Turkey have the most suitable condition for pomegranate growing. Hicaznar, Silifke abysy, Katyrbaby, Canernar, Yzmir 1513, Yzmir 23, Pomegranate cultivars are popular in Turkey. Turkey has rich pomegranate genetic resources and there are different types and forms. Alata Horticulture Research Institute (mersin) has a collection consists of 270 accessions. Cukurova University Agriculture faculty, Horticulture Department(Adana), Aegean University, Agriculture faculty Horticulture Dept (Yzmir) Agean Agriculture Institute (Yzmir). West Mediterranean Agricultural Research Institute (Antalya) has also pomegranate collections. Pomegranate accessions were selected from mainly Mediterranean and Aegean Regions and other location of Turkey. These accessions are examined according to morphological and pomological characters. Pomegranate genotypes have a wide diversity for pomological traits.

Bodaghi *et. al.,*(2009) investigated qualitative and quantitative characteristics of five export pomegranate cultivars in Saveh region of Iran : Pomegranate is one of the important orchard products in Iran and has different cultivars. This study was carried out in Saveh research site for investigation of qualitative and quantitative characteristics of five cultivars (malassaveh, malas yazdi, rabab, naderi, shishe kap). The investigation used a completely randomized design with 5 treatments, 3 replications and the main characteristics were flower number, percentage infection to *Spectrobates cerationiae*, soluble solids and fruit sugar. Following are the results obtained Cv. Malas Saveh had highest amount of vitamin C among Malas Saveh. The fruit weight, diameter and total soluble solids, in fruit was highest among cultivars and is classified as large fruits Cv. Shishe kap had highest number of fruit, flower and highest amount of sugar and percentage fruit cracking in comparison with the other cultivars Cv. Rabab had lowest degree of infection to *Spectrobates ceratoniae* among cultivars that were under investigation; Cv. Naderi had the tallest length in comparison with other cultivars. All five cultivars are cultivated commercially and have red skin and seed. It could be said that the best cultivars are Malas Saveh and Rabab because of vitamin C content and shiny skin.

Jayant kumar *et. al.,*(2009) in their study in mid hills of Himachal Pradesh, India, the performance of sixteen pomegranate cultivars namely Achikdana, Anarshirin, Mohammad Ali, Chawla, Ganesh, G-137, Jalore seedless, Jodhpur Red, Kandhari Hansi, Kandhari Kabuli, Nabha, P-75 - K-3, P-75-K-5, P-23, P-26, Mirdula and Bhagwa was studied. The average highest yield for the last five years were recorded with Chawla (24.3 kg/tree) followed by Nabha (22.7 kg/tree) and Kandhari Kabuli (21.3 kg/tree) and the lowest with Jodhpur Red (15.6 kg/tree). The highest fruit weight was recorded with P-75-K-5 (450 g) and the lowest with Nabha (320 g). The highest total soluble solid was recorded with Kandhari Kabuli and lowest with Nabha. The study of general characters revealed that Kandhari Kabuli and

Bhagwa were the best cultivars. Kandhari Kabuli has pinkish red fruit with red and soft grain and sweet taste. These cultivars were found to be the best for cultivation in the mid hill zone of Himachal Pradesh, India.

Kezban Yezici *et. al.,*(2009) studied the anatomy of pomegranate (*Punica granatum L*) Cv. Hicaznar fruit exocarp. Pomegranate is an important crop with an increasing production for fresh consumption, juice production and medicinal purpose. Pomegranate has become a popular fruit crop and research related with this fruit crop recently intensified significantly. In this study anatomical differences in the exocarp of pomegranate fruits were investigated. Present study was conducted in West Mediterranean Research Institute using Hiczanar, the most important pomegranate cultivar produced and exported from this region. Standard protocols (with slight modification) were followed to study the fruit surface anatomy. As a result it was determined that there is a cuticle layer on the outer surface (pericarps) of pomegranate fruits followed by epidermal cells organized by one layer immediately underneath the cuticle. Epidermis forms the outermost cell layer in the structure of pomegranate fruit peel. As in some fruit species, lenticles, which function as stomata were observed on the pomegranate fruit. Lenticles found in the peel of the pomegranate fruit were evenly distributed on the epidermis. The observed lenticles were mostly in the shape of lens and characterized as extending from loosely lined complementary cell surface. In some fruits gas exchange is performed by lenticles on the surface of fruit peel. Lenticles found in the epidermal cell layer of pomegranate fruit peel may also function the same way. In addition, it was determined that parenchyma cells on the peel of pomegranate surface are in the isodiameteric shape and vascular bundles and sclerenchyma cells. İt was found that protective sclerenchyma cells were densely distributed on the peel of the fruit. This study present anatomic observation of structure of healthy pomegranate fruit peel and it will form the basis for further studies related with abnormal peel structure caused by different physiological pathological and structural problems.

Prerak Bhatanagar *et. al.,*(2009) studied Nutritional Survey of pomegranate fruit Cv. Ganesh in arid soil of western Rajasthan. The average fruit weight ranged from 129.79 to 88.03 g where as fruit length varied from 6.84 cm to 5.65 cm. the aril percentage on whole fruit basis varied from 57.91 to 63.72 percent. Total soluble solids content varied from 10.42 to 12.00 ^{0} brix and the total titrable acidity ranged between 0.45 to 0.51 percent. The Tss: Acidity ratio varied between 20.43 to 26.66:1. Ascorbic acid content varied between 9.75 mg/100 ml and 20.00 mg/100 ml. Total sugars varied from 8.32 to 8.95 percent and reducing sugar varied from 3.47 to 3.61 percent. The nutritional survey of Cv. Ganesh in Bikaner district revealed that immense scope and potential of pomegranate in irrigated arid soils of western Rajasthan. The fruit quality of pomegranate Cv. Ganesh can further be improved by adopting suitable recommended doses of manures and fertilizers and irrigation scheduling.

Dinesh *et. al.,*(2009) studied the flowering biology of pomegranate cultivar 'Ganesh' under Solapur condition of Maharashtra. A preliminary study on pomegranate envisaged as the tree of life. As in depth information on flower biology is an indispensable pre-requisite for complete comprehension of crop improvement. Pomegranate is capable of producing profused flower throughout the year. The flower biology vary with variety, weather, place, period of study. The flowering takes place in three season i.e., Ambebahar (Jan-Feb), Mrigbahar (June-July), Hastabahar (Sept-Oct) in Ganesh cultivar. The observations were recorded with respect to flower bud initiation commencement, cessation, duration of flowering, period of anthesis and its peak time dehiscence. The weight of the mature ready to crack flower bud ranged from 2.2 to 2.8 g per flower. The time taken for flower bud differentiation ranged from 19.9 to 23.5 days. The mean weight of hermaphrodite, intermediate and male flower was found to be 4.82, 2.62, 1.7 g respectively. The hermaphrodite flower featured the miraculous phenomenon heterostyled. The pistil length of hermaphrodite, intermediate and male flower were found to be 2.00, 1.55, 0.65 cm respectively. The pollen viability with acetocarmine stain ranged from 85 to 94%. The flowers were selfed and the fruit set was found to be in the range of 16.66 to 31.57%. In this investigation the other aspects of flower biology are also mirrored in depth.

Babu *et. al.,*(2009) studied the blossom biology of pomegranate cultivar Bhagwa under semi arid tropics of western India. The pomegranate multi seeded apple is an important crop of arid and semi-arid regions of the world. The knowledge on blossom biology provides pre hand idea on the breeding mechanism. Pomegranate a versatile crop with high degree of amenability for flower regulation. The basic particulars pertaining to various aspects of flowering biology viz., flower bud initiation, commencement, cessation, duration of flowering, period of anthesis, its peak time etc, were studied in cultivar Bhagwa under semi arid condition of Solapur during 2008-09. Ambebahar (Jan-Feb), Mrigbahar (June-July), Hastabahar (Sept-Oct). The weight of the mature ready to crack bud flower range from 2.0 to 2.6 g. The time taken for flower bud differentiation ranged from 18.4 to 22.0 days. The weight of the hermaphrodite, intermediate and male flower was found to be 3.75, 2.61, 1.48 g respectively. The hermaphrodite, intermediate and male flowers was measured to be 1.96, 1.45, 0.25 cm respectively. The pollen viability with acetocarmine stain ranged from 84-95%. The fruit set under selfing was found to be in the range of 16 to 20%. The attributes of calyx, corolla, androecium and gynoecium were also studied in the investigation.

Feryal Varasteh *et. al.,*(2009) studied the palynological evaluation of pomegranate cultivar using scanning electron microscopy. In pomegranate pollen grains due to their special features and taxonomy to distinguish and identify the possible difference between species, trees within species. There are many wild and cultivated pomegranate throughout Iran, in the present study pollen grain of some pomegranate cultivars were examined using scanning electron microscopy (SEM). Pollen grains were observed and photographed at 3000′ for whole grain and at 20,000′ for exine patterns. The polar (P) and equatorial axis (E) of the grains, P/E ratio, distance between the two furrows and width of the pollen grain ridges were measured. The pollen grains of pomegranate cultivars showed diversity in size and were placed in prolate (elliptical) shape group based on the P/E ratio. In addition all studied cultivars showed similar foveolate ornamentation pattern, pomegranate pollen morphology descriptive analysis using SEM offers a useful tool for germplasm identification along with other research tools like molecular markers.

Meena, (2009) studied evaluation of pomegranate genotypes for morphological and flowering characteristics under semi-arid climate. Twenty four diverse bearing 8 to 10 year old pomegranate genotypes both exotic and indigenous were evaluated for morphological and flowering characteristics. There were 10 genotypes viz., Alandi, Kandhari, Jalore seedless, Ganesh, P-23, P-26, Bassein seedless, Dholka, Jyoti, Jodhpur Red and G-137 as well as 14 exotic genotype viz., Siah shirin, Achik dana, Gule-shah, Kazak Anar, Sur sakkar, Speen Danedar, Speen sakarin, Kali shirin, Muscat, AK Anar, Khoj, Bedna sadana shirin Anar. The observation recorded were plant height, plant spread, nature of the plant height (3.2 m), E – W plant spread (2.63 m) and stomata density (130.67 mm^{-2}). Minimum days for flowering was recorded in Sur sakkar (16.00) where as maximum flowering duration was in Dholka (39 days) followed by Bedana sadana and Muscat (38.00) with overall population mean of 26.91 days. Duration of dormancy ranged from 56.00 days in Sur sakkar to 67 in Gul-e-shah followed by Kazak Anar and Kalishirin (66.00). Out of 24 genotypes only 12 were found to be dormant type. Seven genotypes were noted to be of spreading type while remaining eight and nine were semi-spreading and erect type respectively. Speen Danedar under Delhi conditions recorded lowest yield (4.75 kg/plant) while Kandhari produced highest yield 11.60 kg/plant.

Mir *et. al.,*(2009) studied the pomegranate cultivars under temperate region in Kashmir valley, the results showed significant variation among all the characters under test. The maximum plant height and spread was noticed in Cv. Kabul Kandhari, where as maximum number of branches/plant was recorded in Cv. Jyoti. Maximum duration of flowering and days taken from flower opening to fruitlet development was noticed in Cv. Jyoti. The highest fruit set, number of fruit/plant and gross yield/plant was observed in Cv. Dholka. Highest fruit weight, diameter and volume were observed in Cv. Bedana. The lowest rind weight was observed in Cv. Mirdula while Dholka recorded the highest aril weight.

Pal and Singh, (2009) studied the association between flowering and yield characters in seven cultivars of pomegranate viz., (Russian Safed Dana, Russian Lal Dana, Kali shirin, Japanese seedling meetha, Ganesh, Kandhari and Dholka) collected from diverse climatic zone. Fourteen characters were recorded round the year and correlations (ã) among those parameters were worked out. It was observed that association of number of floral buds was highly significant with number of solitary, clustered buds dropped, flowers opened and number of flowers dropped and significant with number of fruits dropped, size of fruit and fruits retained. Number of clustered bud was positively correlated with all the characters except days taken to anthesis. Days taken to anthesis showed negative and non significant association with all the characters. Number of opened flowers was positively and strongly correlated with number of fruit set, fruit dropped, fruit retained and size of fruit. This association indicates that more the number of buds produced, higher will be the number of fruit retained along with number of fruit and flowers dropped. The selection may be effective for the improvement of pomegranate.

Singh *et. al.,* (2009) studied the introduction and evaluation of anardana type pomegranate under hot arid condition in Rajasthan. Pomegranate is used for fresh consumption and processing. The sour type pomegranate is also in demand for garnishing many dishes. Fourteen varieties from Himachal Pradesh foot hills collected evaluated for growth, flowering, fruiting, yield quality of fruits at Central Institute of Arid Horticulture at Bikaner. They were planted at 4 x 6 meter distance. Out of collections, 14 germplasm adopted well to hot climate. The maximum plant height 2.70 m and canopy 2.5 x 2.6 m was recorded in collection No.7 after the sixth year of planting. Flowering and fruiting in all types started after 3-4 years of planting. The flowering during the month of August – September was found better for fruiting and yield. Pests and disease pressure was low under arid environment. At initial state fruiting in germplasm varied from 10 to 40 fruits / plant with a fruit weight of less than 100 g. The quality parameters of fruits also varied among collections.

Chharia *et. al.,*(2009) studied Jalore seedless a evergreen and frost resistant cultivar of pomegranate in Haryana state of India. Ripe pomegranate fruits are consumed fresh, surplus fruits can be utilized in juice extraction, in making syrup and other products like anardana which are dried seeds along with aril. Fruits are nutritious and have medicinal value. They studied two types of cultivar i.e., existing deciduous cultivars Ganesh and Jodhpur Red and newly introduced evergreen cultivar Jalore seedless. In Haryana cultivation of pomegranate is limited because plant shed their leaves in winter and as a result there is heavy flesh of flowering in early spring resulting in crop in rainy season which results in spoilage of fruits. Under such circumstances unripe fruits have to be harvested before onset of monsoon when prices in market are the lowest. However crop regulation in these deciduous cvs by retaining late summer flowers on grown up plants is possible and ripe fruit can be harvested in November – December. But there is no chance of taking spring crop in deciduous cultivars because there is hundred percent leaf fall in winter resulting in spoilage of fruits if any on leafless plants in winter. In the study an attempt had been made to study deciduous nature of different cultivar of pomegranate namely Ganesh, Jalore seedless, Jodhpur Red. It was observed that in mild winters, cultivar Jalore seedless did not shed leaves in January and remained evergreen which was helpful in retaining fruits produced as a result of post rainy season (in September) flowering in this cultivar resulting in harvesting of ripened fruits in spring when market prices were highest. Moreover by retaining fruits longer on evergreen plants in spring, early heavy flush of flowering just after winter was kept under control and was delayed till early summer in cultivar Jalore seedless and as a result winter or spring crops were successfully harvested. However in winter there was a severe frost and temperature in January dropped below freezing points and remained low continuously up to a longer duration. As a result there was heavy leaf fall of Jalore seedless plants in winter resulting in frost damage to fruits on leafless plants and there was cent percent spoilage of fruits in January. Such situation in Jalore seedless occurs occasionally but not always in Haryana.

Singh, (2009) studied the characterization, evaluation, conservation of germplasm (*Punica granatum*) germplasm under Indian Arid ecosystem. In this study to select superior genotypes of pomegranate suitable for arid ecosystem of India, thirty accessions/genotypes collected from various sources and propagated vegetatively were planted in spacing of 5 x 5 m at Central Institute of Arid Horticulture, Bikaner, India. The planted genotypes after six year were evaluated for morphological, growth and yield attributing characteristics like tree height, tree spread, suckering, girth, length of leaf, width, leaf tip, leaf base, leaf shape, thorn length, size of flower, Tss acidity percentage, juice, fruit cracking, seed hardness, juice flavour and aril colour. Generally pomegranate cultivars showed good adoption to the specific soil and climate conditions of arid region and great variability was found among them. The cultivar khog had the maximum plant height (2.70 m). A. K. Anar had the largest canopy (2.90 cm across row and 3.1 cm along row). Wide range of variability was recorded in case of length of thorn, ranging from 7.8 cm in G-137 to 3.6 cm in Kabul. In cultivar GKVK, Speen Danedar, Bocheka lines, Dosetta Malus, Tujetis EC – 104347, fruits were found flat round where as rest of varieties produced fruits of round shape. Similarly there was lot of differences and variation in fruit skin colour, which range from Red (P-26, Mirdula, G-137, Khog, Yeracud, A.K.Anar, Dorsetta Malus, Tujetis EC-104347) to fruits with yellowish red ting (Jalore seedless, Ganesh, Bedanasuri, Jodhpur Red, GKVK-1, Dholka, Coimbator white, P-26, Muscat, Patna -5, Bedana

Thin Skin, Speen Danedar, Bassein seedless. Kazaki Anar, Gule-shah, Sirin Anar, Tabest, Kali sirin) and light red (kurvi-2 EL – 24685, Bocheka lines). Maximum length of fruit (8.7 cm) was recorded in Jalore seedless. Similarly maximum fruit width 7.7 cm was recorded for Dholka. Average fruit weight was maximum (196.0 g) in Dorsetta Malus. Number of fruit per tree ranged from 130 in Jodhpur red to 10 in Patna-5 and Dorsetta Malus, number of arils/fruit ranged from 298 in Patna-5 to 70.0 in Bocheka lines. Maximum fruit cracking (70%) was found in Jodhpur Red and minimum in Kabul and Kazaki Anar (10.0%). Patna -5 (11.00), Gul-e-shah (12.0%), Dorsetta Malus and Jalore seedless (15.0%). Maximum juice per cent (52.0%) was recorded in Mirdula Jalore seedless, (51.0%) P-26. Maximum Tss (21.0 °B) was recorded in A. K. Anar, Speen Danedar (20.5 °B). Acidity was recorded in the range of maximum (3.7%) Tujetis EC-104347, Dorsetta Malus (3.41%), Sirin Anar (2.3%), Kali sirin (2.88%) whereas minimum was recorded in Bedana Thin Skin (0.68%).

Pachankar *et. al.,*(2009) studies the numerical taxonomic study of pomegranate. Cluster analysis and principal component analysis, based on 96 characters were applied to 59 cultivars of (*Punica granatum L*). The correlation dendrogram showed less structure than distance based dendrogram. There were more OUTs which do not belong to any group at smaller distance. Principle component analysis (PCA) was found useful and informative. The cultivar Yeracud and Yeracud HRS were found almost identical as these showed maximum affinities while the cultivars A. K. Anar and Alandi were found to be the extreme types. The cluster analysis using correlation coefficient gave poorer overall than cluster analysis using distance coefficients. This technique therefore seems to be useful tool for classifying the pomegranate cultivars in a natural way.

Pachankar *et. al.,*(2009) studied a key for identification of pomegranate cultivars where 12 cultivars of pomegranate viz., Bassein seedless, Dholka, Ganesh, G-137, Jalore seedless, Jodhpur Red, Jyoti, Mirdula, Muskat, P-23, P-26, Yeracud local based on morphological characters were prepared. Qualitative characters being more stable and dependable than quantitative one have been given importance. Seed hardness were most important characters used as primary and secondary characters respectively, colour of various organs was useful in preparation of a key.

Vahid Akbarpour *et. al.,*(2009) studied that physical and chemical properties of Iranian pomegranate in the maturation stage. Pomegranate fruits from varieties Rabbab, Malas – e – Yazd, Malas – e – Saveh, Shishe – kap. Khazare Bardeskan, Naderi, Alak, Abdandan, Tabrizi, Poost syah – e – Saveh, Poost syah – e – Badrood and Lamsari – e – Behshahr were used in the study. Physio-chemical studies of the varieties included fruit fresh weight and volume, weight density, fruit neck diameter, fruit peel weight, fruit peel percent, juice content, peel thickness, seed percent, total soluble solids, titrable acidity, harvest index, reducing sugars, pH, Electric conductivity (EC), Vitamin C, Ellagic acid tannin content of juice and peel (using HPLC), total antioxidants activity of juice and peel (using FRAP assay). The data were applied in completely randomized design, using SAS software and least significant difference range test. Results showed that there were significant differences among varieties which indicated the variation among pomegranate varieties of Iran.

Sephavand *et. al.,*(2009) studied the phenotype characterization of Ten pomegranate cultivars at Karaj condition of Iran introduced from Saveh. The largest fruit produced were poost – sefide – Torsh cultivar, while the smallest fruits were for Alake and Siahe Shirin cultivars. Difference between cultivars for fruit cracking as well as their susceptibility to sunburn and other characters were also detected. According to results Karaj with mean annual temperature of 13.6°C was found not suitable for commercial pomegranate cultivation due to its winter (that during some years goes to lower than – 12°C) thus posing severe hazards.

Mahmoodi *et. al.,*(2009) studied the relationship among fruit quantitative and qualitative characteristics of eight pomegranate genotypes in Ferdows region of Iran. Studies on genotype characteristics generate useful information for breeding programmes. The most important qualitative and quantitative characters of pomegranate seed and fruit were studied for categorizing 8 pomegranate genotypes in Ferdows region (located in Khorasan razavi province). In the study 16 qualitative and quantitative characters of pomegranate seed and fruit (e.g. total soluble solids, titrable acidity, anthocyanin, pH, 100 aril fresh weight and dry weight, length and diameter of aril and seed, peel thickness, fruits crown length and neck diameter etc were evaluated. Analysis of variance showed that all characteristics examined genotypes were significant, showing high variability for these characteristics. Results of mean comparison of characteristics showed qualification of quantitative and qualitative characteristics of various fruit genotypes that were used for judging the utilization type and breeding plans. Results of mean comparison showed that Shahvar genotypes has large aril, low peel thickness, low peel percentage and low flavour index hence find suitable for processing. Shishe kabe Ferdows genotype has high peel thickness, high peel percentage, high and thin crown and found suitable for long time storage. Cluster analysis was performed using sixteen characteristics, in distance of 15 and they are divided in to 5 main clusters. This division showed relation of genotypes that could be used for breeding plans.

Bodaghi *et. al.,*(2009) took comparative studies of fruit characteristics of five export pomegranate cultivars in Saveh region of Iran. Pomegranate is also cultivated on the margin of Kavir with hot summer and cold winters, relatively saline area. The study was conducted on 5 cultivars Malas saveh, Malas yazdi,

Rabab, Naderi, Shishe kap). The completely randomized design with 5 treatments and 3 replications was used to study main characteristics including flower numbers, fruit numbers, percentage of infection to *S. cerationiae*, percentage of fruit cracking fruit weight, fruit length, fruit diameter, ten seed weight, vitamin C content, pH, total acidity, total soluble solids and fruit sugar. The range of percentage of infection to *S. cerationiae*, the percentage of fruit cracking and ten seed weight are 4.9 – 33.62%, 0.95 – 13.96% and 3.4 – 5.1 g respectively. The average of vitamin C content, pH, total acidity, total soluble solids, fruit sugar content were in the range of 8.8 – 22.8 mg/100 ml, 3.5 – 3.7, 1.1 – 1.3, 4.4 – 5.08, 1.6 – 3.5 g/100 cc respectively.

Mars messaoud and Faten Zaouay (2009) studied the agronomic behaviour and diversity of Tunisian pomegranate (*Punica granatum*) cultivars. The aim of this research was to characterize and to study the agronomic behaviour of 44 Tunisian Pomegranate cultivars during three consecutive years all cultivars were examined for their fruit and juice characters. Significant differences were observed between cultivars. Some were highly productive as those named considering all the evaluated parameters it may be stated that the best cultivars from an agronomic point of view are Jebali 3 had the biggest fruit size and juice yield. Gabsi 19 which presented a high full size and nice pink skin and Beyounsi 1 that yielded the highest juice content. Among the sour cultivars Mezzi 3 produced big fruits and highest Total soluble solids content (16.3 °B) with highest pH value (3.6) while Mezzi 1 showed considerable lower juice pH (2.7) but an agreeable red juice. Significant cultivars x year interaction were found for all traits reflecting the climate effect on fruit quality. Phenotypic similarities and dissimilarities were noted between some accessions indicating cases of misidentification. Some cultivars were individualized based on particular characters such as Jebali 3, Garoussi 1, Beyounsi 1; others grouped together and may be considered as polyclonal varieties are Zehri 1, Zehri 2, Zehri 3, Zehri 4, Zehri 5, Zehri 6. These results are useful for pomegranate clonal selection and genetic resource management.

Chapter 6

Training and Pruning

To avoid damage due to stem borer, a multi stem training with 4-5 stems per hill was found beneficial. Removal of water shoots, cross branches, dead and diseased portions and suckers is necessary. Pruning of old spurs (short branches) to encourage growth of new ones is suggested. Some useful tips on pomegranate pruning are:

1) Fruitful and differentiated buds are located at the distal portion of the branches,
2) Pruning of terminal portion of branch would lower down the total flower production,
3) Pruning does not affect sex ratio and fruit quality in pomegranate,
4) Pruning significantly affects total fruits, marketable and unmarketable fruits. Fruit size and yield of higher grade fruits is more with intensity of pruning.
5) Pruning minimizes the bending of branches and also the staking. Based on estimated money returns, 40 cm pruning of main stems was the best.

Sheikh and Rao (2006) studied the effect of pruning and fruit load on yield and quality in pomegranate. Severe pruning and retaining 30 fruit load per plant resulted in recording highest average fruit weight (424.28 g) which amounted to 67 percent rise as compared to unpruned. This treatment however reduced the over all yield of the tree by about 20 percent. It is possible that higher average fruit weight observed in the pruned tree may fetch higher price in the market, thus compensating loss in the yield with regard to quality of the fruit, the fruit developed on the tree which were subjected to severe pruning with 30 fruit load per tree were superior in respect of 100 aril weight, total soluble solid and total sugar content in two year old pomegranate tree.

Gill *et. al.,* (2009) studied the influence of training systems on growth, yield and fruit quality of pomegranate Cv. Kandhari. A ideal training system which gives good quality fruits. The plants were planted in 4 x 4 m, bush, 15, 30, 40 cm trunk height from ground level; 4 to 5 branches from ground level were allowed in all directions were retained as bush. In other training system no shoots were allowed from ground level. Data was recorded for growth, fruit number, fruit yield, size, Tss, acidity, peel, rag, grain, juice percentage. Plants trained to bush system had maximum plant spread and canopy area. Taller plants were observed in 30 cm trunk height trained plants. The average number of fruit decreased with increase stem height of plants and yield also follow the same trend. Bush trained plants also produced large size fruits and smallest fruits were observed in 45 cm trunk height trained plants. Highest Tss and Juice percentage was recorded in fruits harvested from bush trained trees. It could be concluded that bush trained plants improves the productivity of pomegranate and offers the possibility to obtain high quality fruits.

Masalkar *et. al.,* (2009) studied the effect of pruning levels on yield and quality of pomegranate. The yield and quality depends upon management of bahar treatment, includes selection of bahar, percentage of leaf shedding after the rest period, pruning method, intensity of pruning, application of organic and inorganic fertilizers, use of micronutrients, water management, control of pest and disease. Pruning improves fruit quality and minimizes the pest and disease population. Seven pruning levels were tried. Maximum plant spread was in without pruning (2.41 m^2) and was at par with treatment i.e., 30 cm pruning with thinning (2.36 cm^2), pruning 60 cm with thinning (2.22 cm^2), pruning 30 cm without thinning (2.33 m^2) and pruning 20 cm pruning with thinning while it was late 47.07 days in without pruning. The days required for harvesting were at par with each other in all treatments. Average number of fruits registered maximum (80.80) in the treatment of pruning 20 cm without pruning and was at par with remaining treatments. The maximum fruit yield 19.76 kg/tree was observed in treatment 20 cm pruning with thinning and was at par with all treatments.

Chapter 7

Fertiliser Management

Even though pomegranate grows well in soils of low fertility, the productivity can be increased greatly by adequate application of manures and fertilizers. If these are applied based on soil analysis best results in terms of yield and quality could be achieved. The doses of these largely depend upon soil type, age of the trees and crop load and also time of application.

For young plants, 2-3 months after planting, 200 gm of Neem Cake along with 3-4 kg of well rotten FYM has to be applied for initial good start. Three months after this, application of 250 gm DAP, 250 gm Neem Cake along with 5 kg FYM is recommended, Again after 9 months each plant is to be applied with 500 gm DAP, 250 gm sulphate of potash, 1 kg neem cake and 10 kg well rotten FYM

The following doses of fertilizers have been recommended for Rahuri region of Maharastra State.

Age (Years)	Dose (g/Tree)		
	N	P_2O_5	K_2O
1 ½ - 2	250	125	125
2 ½ - 3	500	125	125
3 ½ - 5	500	125	125
6 and above	625	250	250

Along with the above fertilizers 10 kg of FYM is to be applied in first year and the quantity may be increased to about 30 kg from fifth year onwards. These manures and fertilizers are to be applied at the time of Bahar treatment and at different stages of fruit development.

The above fertilizer schedule may be adopted in our state till the research data is made available on the fertilizer schedule. When the fruit is developing water soluble fertilizers are given through fertigation based on leaf analysis and the size and quality of fruits can be improved as per the export needs.

In drought prone areas minor nutrient deficiencies appear frequently and interfere with fruit quality. Zinc deficiency leads to reduction of size in young leaves and margin curling. To correct deficiency of zinc 1-2 sprays of zinc sulphate at 0.6

percent is recommended. Cracking of young fruits will occur even though the water management is proper, when there is boron deficiency. It can be corrected by spraying borax @ 1 g L at 15-20 days interval or alternatively by soil application of borax @ 12.5 g per tree at the time of fertilizer application.

For correcting the nutritional deficiencies commercial formulations such as Macrolic @ 0.33 % at the time of fruit growth followed by 1 % spray 15-20 days after the first spray results in increased fruit size, yield and quality. Similarly, composite nutrient sprays of 0.4 % Ferrous sulphate + 0.3 % Manganese sulphate + 0.2 % Boric acid + 0 . 3 % Zinc sulphate if sprayed before flowering, during flowering and fruit setting results in increased yield and quality of fruits. Use of biofertilizers like Azatobactor reduces the requirement of nitrogen by 2/3rd .

SAMPLING TECHNIQUE FOR LEAF ANALYSIS:

It was suggested that 8th pair of leaf (from growing tip) from new growth in April for Ambebahar crop and in August for Mrigbahar crop has to be collected. It should be from a vegetative terminal. Further it is suggested that composite sample of 50 leaves has to be collected from North, South, East and West directions.

LEAF NUTRIENT STANDARDS:

Leaf nutrient ranges vary for satisfactory performance of pomegranate cultivars. The optimum ranges for N, P and K are 1.15 to 2.5%, 0.13 to 0.39% and 0.124 to 0.9 %respectively. The variation in the range may be due to the age of the trees, beside other factors.

MAJOR NUTRIENTS

Nitrogen

Nitrogen in pomegranate helps in increasing juice percentage, rind percentage and titrable acidity. Total soluble solids, TSS/acidity ratio and ascorbic acid content are also influenced by nitrogen.

Deficiency Symptoms: - The symptoms are first noticed in old leaves and later in younger ones. The leaves become uniformly chlorotic yellow with light green midribs and veins. In extreme cases, necrotic spots develop on the chlorotic margins of the leaves. The overall growth of plant is checked and few thin branches develop. The leaf size is reduced and defoliation occurs. The yield reduced considerably with small fruits (Wavhal, 1993).

Phosphorus

Deficiency symptoms: - The suppression of overall growth is observed. The shoots and stems become thin, number of branches and leaves are reduced; leaves of middle and apical portion of tertiary and side shoots become chlorotic yellow and the basal older leaves remain dull green and lose luster. The chlorosis starts at margins and tips of the leaves and the margins of young leaves roll (Wavhal, *et. al.*, 1985).

Potassium

Deficiency symptoms: - The growth of plant gets reduced producing lanky shoots. The size and number of leaves are reduced. The leaves at the middle position on the shoot develop chlorotic symptoms on the margins with low patches on the lamina having rusty brown or black grey spots. The backward curling of leaves from tips to base is seen (Wavhal, *et. al*, 1985).

Correction :- Potassium fertilizers such as potassium sulphate or potassium chloride can be used.

SECONDARY NUTRIENTS

Calcium

Deficiency symptoms:- Calcium being a relatively immobile element, deficiency symptoms appear in young leaves. Reduction in growth of meristamatic tissue, in young leaves and growing tips is an early indication. Calcium deficient leaves later become malformed and sizes reduced. This often leads to leaf tip burn. It also affects fruits. Fruits become small in size and early maturity will take place.

Correction: - Spraying of calcium nitrate (0.2 %) on foliage after fruit set can be given.

UNIQUE BENEFITS OF CALCIUM NITRATE (CN)

- Calcium improves strength of cell wall, promotes root and shoot growth.
- CN makes plant healthy and tolerant to fungal and bacterial diseases, water stress, etc.
- CN correct Calcium deficiencies/disorders most effectively.
- Hidden deficiency of Calcium, resulting in poor quality of produce is a common problem in many high value crops. The only solution to prevent this problem is to apply water soluble Calcium as CN.
- CN improves quality and yield of produce.
- CN reduces post harvest losses, improves shelf life and offers better storage properties of vegetables, fruits and flowers.
- The Nitrate in CN is non-acidifying, non-volatile and the preferred form for most crops.
- Presence of Nitrate in CN promotes Calcium uptake and vice-versa.
- The water soluble Calcium and Nitrate in CN are readily available at the root zone for quick uptake resulting in fast growth response of plants.

- Presence of Calcium is low in acidic soils and availability of calcium is poor in alkaline and saline soils. In all these soils, CN works better than other sources like CAN, limestone, dolomite and gypsum.
- CN improves soil structure resulting in better porosity, water infiltration, root development and growth.
- The only fertilizer that provides water soluble Calcium is CN, whereas CAN does not provide water soluble Calcium.
- CN is the best source for all seasons, more so for winter/cool weather conditions.

FIELD QUALITY (For broadcast application)

- Nursery: **Apply 10-15 g/m^2, about 10 days after sowing.**
- Mainfield: **Suitable for basal and top dressing.**

To meet the Calcium need, apply CN at the rate equivalent to 20 to 25 % of recommended rate of nitrogen. Apply balance quantity of nitrogen through other sources.

SOLUTION QUALITY (For fertigation and foliar spray)

- Fertigation: **Use it alone or with other fertilizers (except phosphate and sulphate) as per practice.**

Generally, 0.5 to 1.0 g/L is recommended.

- **Foliar application: Spray 5 to 10 g/ L, 2 to 3 times, at fortnightly interval, during the peak developmental/reproductive phases.**

Magnesium

Deficiency symptoms: - In pomegranate magnesium deficiency is very common because pomegranate is normally grown under heavy soils. Magnesium deficiency in pomegranate is chlorosis mainly due to loss of chlorophyll; leaves are stiff, brittle and twisted. Overall suppression of growth, interveinal chlorosis of basal leaves with green midribs and veins, necrosis of the chlorotic areas and inverted 'V' shape of basal green area of the chlorotic scorched leaf are observed.

Correction: - Magnesium should be applied at the rate of 18-24 kg/ha while maintaining a favorable balance between K and Mg. Spraying with a 0.2 % magnesium sulphate solution in summer or 0.5 % magnesium sulphate in autumn is generally recommended as a prophylactic measure.

Sulphur

Deficiency symptoms:- Deficiencies are same as those for nitrogen. Reduction in leaf size and chlorosis of leaves are seen. There is a pronounced effect on chlorophyll content of the leaves with suppression of growth so that the plants are stunted with few branches and leaves, heavy defoliation occurs at early stages. The younger leaves become chlorotic, which progresses to older leaves. Necrotic spots develop on the margins of the chlorotic leaves.

Correction:- Sulphur containing fertilizers (Ammonium sulphate, super phosphate, and sulphur containing potassium salts) can be included in the fertilizer schedule.

Micronutrients

Micronutrients are beneficial in increasing yield as well as improving certain physical characters of pomegranate fruits.

Zinc

Zinc has a definite role in increasing the yield of pomegranate.

Application: - Zinc can be applied in the form of spray (0.5 percent zinc sulphate) three times at monthly intervals. Soil application @ 45 grams zinc sulphate per tree is recommended.

Iron

It has role in increasing the number and weight of fruits. It has also role in decreasing peel percentage.

Application:- 0.4 per cent spray with ferrous sulphate three times at monthly intervals.

Manganese

Manganese nutrition helps to increase the quality of pomegranate fruits.

Application: - The spray of 0.3 percent manganese sulphate is recommended.

Copper

Chlorosis is a reported symptom for copper deficiency in pomegranate.

Application: - Copper sulphate (0.4%) foliar spray increases the fruit quality and reduces chlorosis.

Boron

Boron improves rind and fruit colour. It also helps in improving percentage of grain and reduction in percentage of peel.

Deficiency symptoms :- In young fruits, fruit cracking occurs.

Corrections :- Soil application as borax (10 g. per plant) or foliar application (0.1% boric acid) are the usual practice of boron application for pomegranate.

UNIQUE FEATURES OF SOLUBOR

- Boron plays vital role in cell elongation, root and shoot development.
- Boron is essential for growing tips (root and shoot), pollen germination and pollen tube growth.
- Boron has a primary role in cell wall biosynthesis and plasma membrane integrity.
- Boron facilitates short and long distance transport of sugars / photosynthates.
- Boron deficiency is common in many vegetables, fruit crops, cereals, cotton, oil seeds, etc. Even hidden deficiency inhibits flowering, seed development, fruit development, etc.
- Seeds from Boron deficient crops have low germination rate with high percentage of abnormal seedlings.
- Solubor effectively controls Boron deficiency in all crops and soils.
- Solubor contains almost two times more Boron than Borax.
- Boron supplied through Solubor has better foliar uptake and distribution within plant system than through Borax and Boric acid.
- Soil availability of Boron decreases with increasing soil pH., Boron deficiency is common in calcareous soils. Solubor is the best choice in all situations.
- Solubor provides quick acting Boron, which improves quality of fruits, vegetables, grains, seeds, etc.
- Solubor is a recognized "organic fertilizer" in USA.
- Superior physical, chemical and biological properties of Solubor are the key factors for its fine quality and high efficiency.

GENERAL RECOMMENDATION:-

Spray Solution Concentration:- **0.1 to 0.15 % (1 to 1.5 g Solubor/L). Higher or lower concentration can also be used depending on the nature of crops and their requirements.**

Number of sprays: **2 to 3 sprays at 15 to 20 days interval at flowering/fruit setting stage.**

Compatibility :- **Compatible with most agrochemicals.**

Suitability for fertigation;- **Highly suitable, better than all other Boron sources.**

Suitability for soil application :- **Suitable for direct application as well as through water or herbicide spray.**

Molybdenum

Deficiency symptoms:- The symptoms are first noticed in old leaves and later in younger ones. The leaves become uniformly chlorotic yellow with light green midrib and veins. In extreme cases, necrotic spots are also seen.

Correction:- 1 to 2 kg/ha of sodium molybdate (corresponding to 400 to 800 g/ha of molybdenum) or equivalent amounts of ammonium molybdate can be used.

On acid soils with serious molybdenum deficiency molybdenum fertilization may even have virtually no effect unless the soil is limed at the same time. 0.1 % sodium molybdate solution can be used for spraying purpose.

EFFECT OF SPILT APPLICATION OF N AND K FERTILIZERS ON GROWTH AND FRUITING IN GANESH POMEGRANATE (1999).

Sl. No.	Fertilizer dose	No. of splits	Time of application
1	400:200:200 of NPK g / plant (RDF)	1 dose	1st week of March
2	400:200:200 of NPK g / plant	2 (at an interval of 2 months)	1st split is applied in 1st week of March 2nd split is applied in 1st week of May.
3	400:200:200 of NPK g/plant	3 (at an interval of 1 ½ month)	1st split is applied in 1st week of March. 2nd split is applied in 3rd week of April. 3rd split is applied in 1st week of June.
4	400:200:200 of NPK g/plant	4 (at an interval of 1 month)	1st split is applied in 1st week of March. 2nd split is applied in 3rd week of April. 3rd split is applied in 1st week of May. 4th split is applied in 1st week of June.

(M.M. Sheikh Ph.D. Work)

Conclusion:-

In overall situation it may be seen that four split application of fertilizer dose of 400:200:200 g NPK per plant (P applied in one split) has resulted in recording higher yield with maximum average fruit weight, juice percentage, TSS, Sugars, Ascorbic acid as compared to lower number of splits or one dose.

PARAMETER	UNIT	STANDARDS
pH	-	6.00 – 8.00
Electrical conductivity	dS/m	0.00 – 1.00
Organic carbon	%	More than 1.00
Available N	ppm	100 – 200
Available P	ppm	50 – 100
Available K	ppm	500 – 800
Available Ca	ppm	500 – 2000
Available Mg	ppm	350 – 500
Available S	ppm	10 – 50
Available Fe	ppm	2.00 – 10.00
Available Mn	ppm	2.00 – 20.00
Available Zn	ppm	1.00 – 5.00
Available Cu	ppm	0.20 – 2.00

ELEMENT	UNIT	STATUS
Nitrogen	%	2.94
Phosphorus	%	0.11
Potassium	%	0.76
Calcium %	%	2.54
Magnesium	%	0.55
Sulphur	%	0.24
Iron	ppm	100
Manganese	ppm	44
Zinc	ppm	50
Copper	ppm	22

Leaf Nutrient Standards for Pomegranate

Nutrient	Observed Range	DRIS NORMS				
		Deficient	Low	Optimum	High	Excess
N (%)	0.40 – 2.20	<0.54	0.54 – 0.90	0.91 – 1.66	1.67 – 2.04	> 2.04
P (%)	0.08 – 0.33	<0.09	0.09 – 0.11	0.12 – 0.18	0.19 0.21	> 0.21
K (%)	0.20 – 2.05	<0.20	0.20 0.60	0.61 – 1.59	1.60 – 2.26	> 2.26
Ca (%)	0.60 – 2.40	<0.13	0.14 – 0.76	0.77 – 2.02	2.03 – 2.65	> 2.65
Mg(%)	0.16 – 0.49	<0.03	0.03 – 0.15	0.16 – 0.42	0.43 – 0.55	>0.55
S (%)	0.04 – 0.70	<0.10	0.10 – 0.15	0.16 – 0.26	0.26 – 0.42	>0.42
Fe (ppm)	25 – 297	<34	34 – 70	71 – 214	215 – 286	> 286
Mn (ppm)	14 – 99	<15	15 – 28	29 – 89	90 – 119	> 119
Zn (ppm)	7 – 44	<8	8 – 13	14 – 72	73 - 94	>94
Cu (ppm)	21 – 86	<7	8 – 28	29 – 72	73 – 94	>94
Yield (t/ha)	87 - 20	<13.7	13.7 - 15.5	15.6 – 18.8	18.9 – 20.6	>20.6

1. Fertigation is highly efficient technology which delivers fertilizers via irrigation systems, ensuring maximum efficiency of both water and dissolved nutrients usage.
2. These fertilizers are completely soluble in water.
3. The solution does not clog emitters tubes pipes and other irrigation system.
4. These are low in salt index and hence do not leach or get fixed in the soil or increase salinity.
5. These are available in ionic form in water solution and hence are easily absorbed by the plant roots.
6. These contribute acidity to water which counter carbonate reactions, thus avoiding fixation of phosphorous with soil calcium.
7. These are chloride-free thus keeping salinity low and controlling negative effect of chlorides on plant metabolism.
8. As these fertilizers contain sulphates, they enhance incorporation of sugar and starch into fruits and storage organs thus improving product quality.
9. The ratio of different nutrients in these fertilizers grades is such that it enhances uptake efficiency of each nutrient as it creates acidic soil solution in the root zone (pH 6-7).
10. Optimum plant nutrition demand of potassium and minimum demand of micronutrients is met by these fertilizers.
11. These are free flowing and easy to handle.
12. Every grain contains same ratio of nutrient as specified.
13. Split applications through every irrigation are possible without increasing labour cost of repeat fertilization as it occurs in conventional fertilizer application.
14. As these fertilizers are highly concentrated and losses on use are minimum the farmer need not carry large inventory. Also. Smaller packages of 100 g. to 5 kg. besides 25 kg normal bags are also available thus reducing inventory costs.
15. Nitrogen is largely available as Ammonium ion thus avoiding Nitrate and Amide losses. The Ammonium ion easily gets converted to available form to plants as per its requirement and is not lost due to evaporation or leaching.
16. In calcareous soils, the continuous supply of P_2O_5 in low quantities minimizes fixation process of P.
17. The micronutrients are supplied in fully chelated form thus making micronutrients easily available to the plants in slightly acidic pH
18. Electric conductivity of solution is low thus facilitating ionization and absorption by plant roots.

19. Water soluble fertilizer mixtures are used in low concentration. A stock solution of 10 % is sufficient which can further be diluted to 100 times. Thus damage to plants due to high concentration of any nutrient is avoided.
20. Fixation, antagonizing, leaching, evaporation losses during application of these fertilizers is minimum. Thus these are ecologically friendly.
21. No harmful effect on micro fauna and flora due to pH of 6 to 6.5. For horticulture and floriculture under green houses, soluble fertilizers can be used and are most effective for high yield and high quality of produce. More and more farmers and cultivators are switching over to use of water soluble fertilizers especially in Maharashtra, Karnataka, Andhra Pradesh and Tamil Nadu. Nearly 3,30,000 hectares of land has come under drip/sprinkler irrigation systems. Water soluble fertilizer mixture requirement calls for nearly 80000 tones per annum to meet the existing drip/sprinkler irrigation area.

Edward Raja (2006) studied the susceptibility of pomegranate cultivars in India to nutrient disorders. Ganesh cultivar susceptible to potash 0.52% leaf K, where as the critical level for K is 1.4% and lower leaves exhibited marginal chlorosis from 30 days to harvest. The K deficiency also affected aril colour esulting in reduced market rate. The cultivar Bhagawa was susceptible to Manganese (Mn) Zinc (Zn) deficiency and the leaf Mn was as low as 12-15 ppm resulting in poor aril colour. This variety has a characteristic of continuous flowering and fruit set affecting nutrient uptake efficiency of roots due to continuous stress on the supply of carbohydrates to roots. The variety Mirdula is susceptible to Manganese and Zinc deficiency. Where as cv. Joythi was susceptible to Mn deficiency even though the soil had high available Mn at 14.3 ppm where as the soil critical level is of Mn is 2.5 ppm only.

Mehdipanahi *et. al.,*(2006) studied the influence of irrigation and mineral nutrients on growth, yield, quantity and quality of pomegranate fruit in Iran. The effect of macro (N, Mg, K) and micronutrients (Fe and Zn) on yield, fruit quality, yield, fruit set, fruit weight, berry number, berry volume, total soluble solids, pH, acidity were influenced by N, K, Mg, Fe, Zn, Potassium application increased fruit yield and weight. Yield was positively correlated to the total soil N at 30 – 60 cm depth. Nitrogen application increased vegetative growth of the main and lateral canes. Soil N did not correlate strongly with tissue N status.

Sheikh and Rao (2006) studied the effect of split application of N and K on growth, fruiting in Ganesh pomegranate. Higher split (4) affected flowering, however the fruit set increased with two split. The highest numbers of fruits, fruit weight were observed in four split. The rind became thicker. Higher Nitrogen and Potassium content were maximum, Tss and Tss: Acid ratio observed in four split.

Dhillon *et. al.,* (2009) studied the effect of Nitrogen, Phosphorus and Potassium on growth, yield quality of pomegranate Cv. Kandhari. Fertilizer is a vital input in fruit production to determine yield, size, and high quality fruits. The treatment consists of four levels of N (0, 20, 40, and 60 g/tree/year) three levels of P_2O_5 (0, 20, 50 g/tree/year) and four levels of K_2O (0, 20, 40, and 60 g/tree/year), while applying a particular variable the fertilizer nutrient, the level of rest of the nutrient was kept constant. The dose of nutrient stabilized after 5 years of age. Nitrogen increased the growth of the plant in terms of higher girth, height, spread and volume of tree as compared to rest of the treatment. The number of fruits per plant largely contributing yield, increased linearly with the increase in nitrogen upto maximum level. Similarly all the P and K treatments also recorded higher yield as compared to yield that obtained when respective elements were omitted, though the increase was not linear. The fruit size increased under all N, P and K treatments. However the trees received no K produced lowest fruit size. Quality analysis of the fruits showed maximum Tss with 40 g N tree year^{-1} treatment. However the acid content was not significantly affected by any of the treatments. Invariably all the N, P, K doses improved the rag as well as grain juice percentage. However higher doses of N,P,K slightly reduced the amount of Juice% but increased seed weight. In overall N2, P1, K2 treatments are showing encouraging results as far as fruit weight, yield, quality is concerned.

Meena *et. al.,* (2009) studied integrated nutrient management in pomegranate Cv. Ganesh. The productivity largely depends on proper application of manures and fertilizers. The 7 treatment comprises of inorganic fertilizer (NPK) organic manures (vermicompost, neem cake, karanj cake) and bio-fertilizers (Azotobacter and PSB). The experiment was conducted in randomized block design with four replications. Two years observation revealed that the maximum increase in shoot length (33.30%) and shoot diameter (28.28%) were recorded at recommended dose of (NPK 500 g N, 200 g P and 500 g K/plot) through inorganic fertilizer at 75 days after treatment. The days required for flowering was not significant among treatments. However, maximum number of flowers (15.35/shoot) and fruit set (45.64%) were recorded at 10 kg vermicompost + 25% recommended dose of NPK + 5 kg neem cake + 20 g PSB/plant treatment. This treatment also exhibited maximum fruit retention (45.73%), fruit length (13.90 cm), diameter (13.85 cm) and fruit weight (250.26 g). As far as quality of fruit is concerned, the highest Tss content (16.95° Brix), Tss : acid ratio (4.62), Ascorbic acid content (14.39/100 ml juice), reducing sugar (12.82%), non reducing sugar (1.35%) and total sugars (14.24%) were recorded at 10 kg vermicompost + 5 kg neem cake + 25% recommended dose of NPK + 20 g PSB/plant treatment compared to other treatments including control. Further this treatment shown highest yield of 36.30 kg/plant as compared to 29.39 kg/plant at recommended dose of NPK through inorganic fertilizer treatment.

Goel *et. al.,*(2009) conducted study on response of application of customized fertilizer grade (CFG) on yield and quality of pomegranate on Inceptisol (typic haplustepts) for validation of CFG. The pomegranate crop was grown with three treatments (T1- Farmers practice, T2-University practice, T3- CFG) in four replications by using randomized block design. Customized fertilizer is a multi-nutrient forms manufactured through a systematic process satisfying the crops nutritional needs, specific to site, soil and stage. The highest marketable yield was recorded in T3 (21.6 tonne/ha^{-1}) followed by T2 (20.6 ton/ha) as compared to T1 (19.6 tonne/ha). The yield difference between T3 and T2 treatments was found significant. Application of customized fertilizer grade N20:P10:K10. S5, Mg^2, Zn05, B0_3: Fe 0.2 @ 1000 kg/ha^{-1}. T3 as basal application to pomegranate recorded significantly more number of large size 'A' grade fruit (32%) as compared to other fruits.

Sadiye *et. al.,* (2009) studied the determination of Macro, Micro nutrient contents of pomegranate Cv. Hicaznar fruits grown in the Western Mediterranean region of Turkey. In this study the aril, seed, peel, juice from pomegranate were evaluated. The macro and mineral content during different growing stages 1st stage: Immature fruits after fruit set, 2nd stage: unripe fruits, 3rd stage: ripe fruit were measured. The percentage of moisture level was lower in the seeds than other parts of the fruits. Nitrogen content was dominant macro element seeds (28.33 – 18.20 g/kg) and in aril (15.7 – 8.5 g/kg). Potassium was shown to be dominant macro element in juice (2.1 – 17.68 g/kg) and peel (10.99 – 12.17 g/kg) respectively. The relative order of concentration of macronutrients both in aril and seeds were N > K > P >Ca >Mg >Na. The relative order of concentration of these macronutrient in juice and peel were different N > K > P >Ca >Mg >Na. The accumulation of all the macro and micro elements within the fruits also increased during fruit growth and development. The concentration of most micronutrient was greater in the unripe fruit than in the ripe fruits except iron. In micro elements iron demonstrated to be the dominant in the seed (60.47 mg/g), aril (30.93 mg/g), (peel 28.06 mg/g) and juice (1.9 mg/g) of ripe fruit respectively. The accumulation of all macro and micro nutrient within the fruit also increased during fruit growth and development. This shows that pomegranate could be a good source of minerals especially with seed consumption.

Ranpise *et. al.,* (2009), An experiment was carried out to observe the effect of chemicals, viz., calcium chloride, calcium nitrate, potassium dihydrogen phosphate with water sprays and control, sprayed with one and two percent concentrations at 90 and 120 days after full anthesis (DAFA). Spraying calcium nitrate (2%) at 90 days after full anthesis significantly influenced the chemical characters of fruits viz., Tss, titrable acidity, ascorbic acid content and it reduced the seed hardness, percentage of seed, incidence of internal breakdown in the fruit. This chemical sprayed at 120 days after full anthesis at two percent also reduced the physiological loss in weight (PLW), rotting acidity and increased the Tss during storage. Application of potassium dihydrogen phosphate (2%) at 90 days after full anthesis influenced the fruit quality in terms of physical characters viz., length, diameter, weight volume, percentage of juice and arils. This chemical applied with two percent concentration at 120 days after full anthesis improved the colour of rind and sugar content of fruit. Hence, it was concluded that pre-harvest spraying of two percent calcium nitrate at 90 days after full anthesis was found beneficial for improving the quality of freshly harvested fruits and at 120 days after full anthesis for improving the storage quality of fruits. The pre-harvest spraying of potassium dihydrogen phosphate with 2% concentration at 120 days DAFA was found to improve the fruit colour.

Kulkarni *et. al.,* (2009) conducted field experiment on effect of different chemical sprays on the fruit yield of pomegranate at Horticulture, Department of MAU, Parbhani, India. The experiment was laid out in randomized block design with three replications and seven treatments viz., K_2SO_4 (0.5%) with 3 sprays, K_2HPO_4 (0.5%) with 3 sprays, K_2SO_4 + K_2HPO_4 (0.5%), cytozyme (0.1%) with 3 sprays, cycocel (500 ppm) with 2 sprays was adopted with control. Spraying of chemicals was recommended at 90 days after fruit set.

Kulkarni *et. al.,* (2009) studied the effect of different chemicals on the fruit yield and quality of pomegranate variety Ganesh with seven treatments viz., K_2SO_4 (0.5%) with 3 spray, KH_2PO_4 (0.5%) with 3 spray, K_2SO_4 + KH_2PO_4 (0.5%) cytozyme (0.1%) with 3 sprays, Cycocel (500 ppm) with 2 spray, Ethrel 100 ppm with single spray was adopted with control spraying of chemicals was commenced at 90 days after fruit set and were undertaken at 15 days interval. The observation on fruit drop, number of fruits harvested maturity period, yield per tree, weight and volume of fruit, weight of aril, juice percentage, Tss titrable acidity and ascorbic acid were recorded. The treatment K_2SO_4 + KH_2PO_4 (0.5%) as well as cycocel (500 ppm) significantly increased number of fruits per tree on the contrary Ethrel treatment reduced the number of fruits per tree than control. Ethrel & Cycocel application significantly reduced the number of days required for maturity of fruits. The treatment KH_2PO_4 and K_2SO_4 (0.5%) + KH_2PO_4 significantly increased the fruit yield. The other treatments namely cytozyme, cycocel, ethrel were found at par with control. The quality of fruits in respect of size and colour can be improved by means of three sprays of KH_2PO_4 (0.5%) at 15 days interval commencing from 90 days of fruit set. For hastening of fruit maturity and colour development two spray of cycocel (500 ppm) at 105 and 120th days of fruit set a single spray of Ethrel 100 ppm at 120 days of fruit set can be recommended.

Nutrition

Nutritional studies (Chougule, 1976; Shende, 1977) indicated that following fertilizer schedule should be given:

Age (year)	FYM (kg)	Gram/tree/year		
		N	P	K
1st Year	10	250	125	125
2nd Year	20	250	125	125
3rd Year	30	500	250	250
4th year onwards	50	625	300	300

The full dose of FYM, P and K and half dose of N should be given at bahar treatment prior to first irrigation. The remaining quantity of N can be given in single dose 40-45 days after first dose. Both macro and micro nutrients affects growth and development in pomegranate. The deficiency symptoms and leaf nutrient standards for macro-nutrients like N, P, K, Ca, Mg and S have been described (Wavhal, 1986) in sand culture studies on foliar applications of B, Mn, Zn and Fe indicated beneficial results. Boron application at 0.2 per cent was found useful to reduce cracking of fruits and improvement of fruit colour. Lal (1975), Pundir and Pathak (1981) and Bankar *et. al.,* (1990) observed that soil application of NPK fertilizer was useful for increasing the production and fruit quality of pomegranate. Lal and Chauhan (1979) reported highest fruit yield with a medium leaf N content (2.35 per cent). Bhargava and Dhander (1987) observed that most of the nutrients were stabilized in 8th pair of leaf from growing tip in pomegranate and may be taken as standard for nutritional diagnosis. While Munde *et. al.,* (1980) reported stable age of leaf for Fe, Mn and Zn as 4, 5 and 11-12 months are respectively.

FERTIGATION

A WHAT IS FERTIGATION?

- An obvious objective of every grower is to enlarge the income from the land he cultivates. The way to achieve this i's to increase crop yield and minimize input costs.
- Among the various factors responsible for high crop yield, the use of proper quantity of fertilizers at appropriate time plays a vital role in enhancing the productivity.
- Fertilizers are normally used as basal dose or top dressing. The full year dose is split into one, two or three doses. Since these fertilizers are applied in bulk, lot of fertilizers go waste due to leaching, volatilization and fixation in the soil. Moreover these fertilizers get transmitted to areas beyond the active root zone and are no longer useful to the plant. In many cases, the effective utilization by the plant is less than 50% of the fertilizer applied.

- Drip irrigation is the most efficient method of irrigation for crops like fruits, flowers and vegetables. This system offers an opportunity for precise application of water soluble fertilizers and other nutrients to the soil at appropriate time with the desired concentration. This application of fertilizers through irrigation water is called fertigation.
- The development of root system is extensive in a restricted volume of soil when cultivation is done, with drip irrigation and application of fertilizer or any other chemical through drip can efficiently place plant nutrients in the zone of highest root concentration. The major advantage of fertigation is that the necessary fertilizers can be applied uniformly to each and every plant even on a daily basis, thereby creating ideal and optimum environment for the plants to absorb required fertilizers and micronutrients.

B. WHY FERTIGATION?

- Fertilizers can be applied timely according to crop requirements at different growth stages.
- Small quantities can be applied at close intervals.
- Irrigation combined with fertilizing maintains a nutrient solution in the root zone. The nutrients are readily available and uptake by the plants is fast and efficient.
- Every plant being irrigated is sure to receive its portion of nutrients.
- Even distribution of nutrients as well as water is maintained in the active root zone.
- Fertigation satisfies the requirements of low volume irrigation (micro sprinklers and Drippers) where root system has a limited volume and hence accurate control of water and nutrients in the root zone is essential.

C. ADVANTAGES OF FERTIGATION

- Improves Availability of nutrients and consequently improved nutrient uptake by roots.
- Application is restricted to wetted area where root activity is most intensive.
- Loss of nutrients by leaching is minimized.
- Soil compacting is prevented.
- Weed population is reduced
- Reduces labour costs
- Lowers the fertilizers and water use, hence saving in input cost.

D. METHODS OF FERTIGATION

Several techniques are available for applying fertilizers through drip irrigation systems. They can be classified into two groups according to their principle of operation.

a) SUCTION SYSTEMS:

1) **Ventury Type** : This is based on the principle of ventury tube. It delivers the fertilizer at a constant concentration which depends on the water flow. The ventury should be connected in parallel with the pipe line.

2) **Fertilizer Pump** : Fertilizer solution is prepared in a tank from which it is pumped and injected into the irrigation system. The fertilizer solution is delivered at a constant concentration. The impeller and casing of such a pump (dozer pump) must be made of non-corrosive material such as nylon or polycarbonate. The advantage of using a pump is that it builds up an additional pressure in the system, unlike a ventury or a fertilizer tank where pressure drop takes place.

b) DELIVERY SYSTEM :

1) **Fertilizer Tank** : A tank containing fertilizer solution is connected to the irrigation pipe at the supply point. Part of the irrigation water is diverted through the tank, diluting the fertilizer solution and returning to the main supply pipe. The concentration of fertilizers in the tank thus becomes gradually reduced with the time.

E. FERTILIZERS AND MICRONUTRIENTS:

1. **NITROGEN:** Nitrogen is most often applied through irrigation systems without any significant clogging problems. Careful consideration must be made of the pH in the irrigation water since some Nitrogen sources, particularly aqueous ammonia and anhydrous ammonia will increase pH. Increased pH can result in precipitation of insoluble calcium and magnesium carbonates that can clog a drip system. Ammonium sulphate should not be used in places where calcium content of water is higher than 70 mg/l.

2. **PHOSPHOROUS** : If we have to apply phosphorous individually through irrigation system, food grade (white) phosphoric acid is recommended. Green phosphoric acid which is most commonly available and used in most of the complex fertilizers manufacturing, contains impurities and precipitates calcium. Maintaining low pH will help in controlling phosphate precipitation.

3. **POTASSIUM:** Potassium is also easily leached in sandy soils and generally must be replenished to maintain a proper N:K ratio for good crop production and quality. 'K' can be injected into the drip irrigation system as potassium chloride or potassium nitrate. These Potassium sources are soluble and have few precipitation problems.

4. **MICRONUTRIENTS:** Generally micronutrients can be applied per plant for most crops and in most soils. If supplied via the drip system, micronutrients such as Iron, Zinc, Copper and Manganese can be injected as chlorides or sulphates. Sulphate salts of micronutrients may react with salts in the irrigation water and cause precipitation and clogging. Chelated micronutrients are highly water soluble and usually cause little clogging or precipitation.

F. ADVANTAGES OF USING 100 % WATER SOLUBLE FERTILIZERS IN FERTIGATION:

- Ensures a regular flow of both water and nutrients resulting in increased growth rates and higher yields, yield increase of more than 100 % is possible to achieve with correct fertigation schedule.
- These fertilizers offer greater versatility in the timing of the nutrient application to meet specific crop demands. Fertilizers can be applied at predetermined times according to the developmental and physiological stages of the crop. It should be noted that delivering the fertilizer in split doses rather than in a single application results in a more pronounced increase in the plant nutrient uptake.
- Improves the availability of nutrients and their uptake by roots.
- The three major nutrients N, P and K supplied in one solution to the plant gives better root absorptions and consequently higher yields.
- Safer application source which eliminates the danger of burning the plants root system, since these fertilizers greatly diluted in irrigation water.
- Offers simpler and more convenient application than any soil application of fertilizer, thus saving time, labour, equipment and energy.
- Insecticides and Herbicides can also be supplied along with these fertilizers through irrigation system since they are compatible with most of them.

G. CATEGORIES OF WATER SOLUBLE FERTILIZERS:

The fertilizers can be classified into two categories basing on the nutrient sources chosen for manufacturing them.

1) **Chloride Free Fertilizers:-** These fertilizers are useful for high value crops and crops which are more sensitive to chlorides such as Tobacco, certain fruits such as Citrus, grapes, arecanut and vegetables such as Tomato, Potato, Cauliflower etc. These fertilizers are produced by using Urea, Ammonium Nitrate, White Phosphoric acid and Potassium Nitrate as basic ingredients. There are a very few (2 or 3) Companies in the world who are supplying this type of high quality fertilizers e.g., Haifa Chemicals Ltd, Haifa, Israel.
2) **Normal Fertilizers:** - These fertilizers are produced by using Ammonium Ni-

trate, Urea, Ammonium Phosphate, Ammonium Sulplate, Green Phosphoric acid, Potassium Chloride and Potassium Sulphate. Since these fertilizers contain chlorides and Sulphates, there is always possibility of clogging of irrigation system due to precipitation of Calcium and Magnesium Bicarbonates, Sodium etc, if irrigation water pH is not maintained. Presence of high concentration of Chlorides in the wet zone will lead to more uptake of Chlorine by plants ultimately leading to shorter shelf life and bad quality of fruits, vegetables, flowers etc. Most of the foreign companies, who have recently entered into India are supplying this category of fertilizers.

H. LIST OF FERTILIZERS NOT TO BE USED FOR FERTIGATION :-

1. Aqueous Ammonia (NH_4OH)
2. Calcium Nitrate
3. Calcium Ammonium Nitrate
4. Potassium Sulphate
5. Zinc Nitrate
6. Ferric Sulphate.

I. SUITABLE FERTILIZER AND NUTRIENT CONCENTRATION FOR FERTIGATION

1. **FERTILIZER CONCENTRATION:** - The concentration of fertilizer in irrigation water can range from 4000 to 10000 ppm. Normally 0.5 % concentration (5000 ppm is maintained for drip irrigation system which is equivalent to 5 g. / litre water. The actual concentration needed depends on the fertilizing material and crop requirement.
2. **NUTRIENT CONCENTRATION:** - For proper growth plant must be supplied with Nitrogen, Phosphorous, Potassium, Calcium, magnesium, Boron, Zinc, Copper, Manganese, Molybdenum, Iron etc. The following table contains the nutrient concentrations which seam best suited for a wide range of crops.

ELEMENT	CONCENTRATION
Nitrate Nitrogen (NO_3)	150-200 ppm
Phosphorous (P_2O_5)	110 ppm (110)
Potassium (K_2O)	240-490 ppm
Calcium (Ca)	150-200 ppm
Magnesium (Mg)	20 ppm
Iron (Fe)	5 ppm
Manganese (Mn)	1 ppm
Copper (Cu)	0.1 ppm
Zinc (Zn)	0.1 ppm
Boron (B)	0.2 ppm
Molybdenum (Mo)	0.05 ppm

Zn – EDTA – 15 %

This product is already covered in FCO and it is widely-used in India. This is a free flowing crystalline powder with 15 % Zinc. In this product the Zinc is chelated by a chelating agent EDTA (Ethylene diamine tetra acetic acid). The pH of the product is 6 to 6.5 with particle size 95 % < 0.5 mm. This product is recommended for soil application as well as for foliar application.

The Zinc deficiencies are always reported on horticultural crops like pomegranate, grapes, mango and other horticultural crops. The zinc EDTA through soil application as well as through foliar application is useful to correct or prevent the Zinc deficiencies. It is useful for uniform sprouting as well as uniform vegetative growth. This product is very well suited for soils pH ranging from 5.0 to 10.0.

Zn EDTA contents 15 % Zinc in which Zinc is 15 % and other is chelating agent and soluble filler material.

Fe – EDTA – 13 % Fe

Fe EDTA contents 13 % Fe and the other is chelating agent. Ethylene diamine tetra acetic acid and other soluble filler material. In India this product is widely used to correct or to prevent the iron deficiencies and iron chlorosis and it is recommended for soil pH ranging from 4 to 8.

This is a free flowing homogeneous product with pH 5.5 to 6.5. The Fe with the chelating agent is readily available to the crops when given through spraying.

Fe (EDDHA) – Fe 6 %

This is the source of Iron. Iron is a micronutrient which is required for plant growth and without which the disorders like iron chlorosis occurs which results in reduction in yields.

Earlier, this nutrient was supplied through product like chelated Iron as Fe EDTA. But for soil application the chelating agent like EDDHA proves better than EDTA.

EDDHA is a chelating agent which is a short form of Ethylene diamine-dihydroxiphenyle acetic acid. Iron is chelated with the chelating agent EDDHA. Fe (EDDHA) contains 6 % Fe with pH value 7-8. This is a homogeneous, micro granulated free flowing product with particle size 95 % < 0.5 mm.

Fe EDDHA is recommended for soil application micro nutrients either through drip or through drenching. For preventive treatment the lower application rates are sufficient while to correct the deficiencies, the higher rates are recommended.

Fe (EDTA), Fe – 12 % is a product which comes under FCO with chelating agent. Ethylene diamine tetra acetic acid. This product is recommended only for the soil pH range upto 8. Whereas this product is not suitable for the pH range more than

8 and above. Considering the limitation in uses of this product in limited pH range recently new chelating agent is invented as EDDHA. This product is recommended on wide pH range from 4.0 to 12.0. Hence, we are importing this product for trail purpose. The application rates are:

	g/plants	
	Initial production	Full production
Horticultural crops	40-60	75-100
Grapes	10-15	15-70
Grapes (wine purpose)	5-7	7-10
Tomato, brinjal etc.	60-80 gm/gunta	
Strawberries	100-150 gm/gunta	
Melons	100-150 gm/gunta	

Fe EDDHA contents 6 % Fe and rest of the material is chelating agent and other soluble fillers.

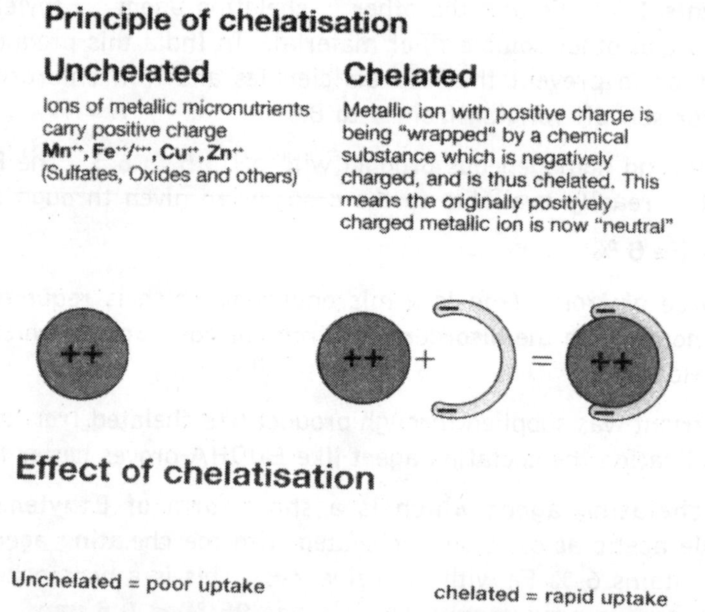

The Water Soluble fertilizer has the following merits over the conventional fertilizers:

1. Soluble fertilizers completely dissolve in water leaving no precipitate. Therefore, there is absolutely no problem in fertigation, clogging of drippers and emitting pipes.
2. Every plants irrigated receives a regular flow of both water and nutrients, directly in the root zone resulting in improving growth and increasing the productivity. Over two fold increase in yield is possible due to correct fertigation schedule.
3. Nutrients can be applied as and when required to meet specific crop nutrient demands. Also fertilizers can be applied at a predetermined time according to the critical growth stages of crops. Frequent fertigation (more split doses) versus single application as in conventional fertilizers results in remarkable plant nutrient uptake which reflects in increased biomass production.
4. Since the nutrients directly reach the rhizosphere in liquid form there is no loss of nutrients especially through volatilization.
5. The fact that three major and micro nutrients are supplied in one solution to plants, it gives better root absorption and consequently higher yields and good quality.
6. Safer application methods eliminate the danger of burning the plant root system, as the fertilizer is greatly diluted in the irrigation water.
7. Since these fertilizers can be applied through drip irrigation through venture, fertilizer tank or injection pump, the application becomes very simple than any soil application of fertilizers and thus results in saving labour, time, energy and overall application cost.
8. Due to frequent application in small doses, the amount of fertilizers present in the soil at any time is small which prevents losses from leaching and run off during heavy rainfall. This results in saving of nutrient requirements per unit area which ultimately results in saving in quantity of fertilizers nearly 25 to 50 % of recommended level.
9. Wide range of compatibility of most soluble fertilizer grades facilities concurrent incorporation of fungicides, pesticides and nematicides, which helps saving in labour, time, energy and its application cost.
10. Since most soluble fertilizers are blended with chelated micro nutrients, there is no necessity to go for additional micro nutrient mixture or sprays which ultimately helps reducing the additional cost on labour and mixtures.
11. Allows crop to be grown on marginal lands, such as sand or rocky soils, where accurate control of water and fertilizers in the plant's root environment is critical.

Water Soluble Fertilizers

N: P: K

1:48:35

Technical Aspects

Analysis

Total Nitrogen N	:	1 %
Nitrate, NO_3-N	:	1 %
Ammonium NH_4-N	:	0 %
Phosphorus (P_2O_5)	:	48 %
Potash (K_2O)	:	35 %

Agronomic Benefits

100% water-soluble fertilizer in a crystalline form. The NPK is especially developed to meet with all the nutrient requirements of the crops, during the second stage after planting.

Increase of yield and better fruit quality. The efficient root uptake of Nitrogen, Phosphorus and Potassium through FERTIGATION guarantee an optimum metabolism stand growth of plants and fruits (formation of nucleic acids, flowering, photosynthesis, perspiration, formation and transport of sugars).

Fertigation

In continuous fertigation, the concentration of the nutrient solution should be between 0.5-1.5 g/l/For the mother solution, a concentration of 10-15 % can be used depending on the water temperature. The number of fertigation applications should be spread in the corresponding period of application. There are several differences between the fertilization of crops grown in greenhouses and those grown in open fields. The uptake of nutrients is higher in protected crops due to faster growth and higher yields. The same applies to crops grown in spring and summer, versus crops grown in winter. The nutrient requirements also depend on climate conditions, soil type, temperature, varieties, methods of application and irrigation.

Use :-

- Prepare a mother solution by adding 15 kg NPK 1:48:35 to 100 L water (20°C). This solution then contains:
- 1.5 gr. N/liter solution
- 72.0 gr. P_2O_5/ liter solution
- 52.5 gr. K_2O/liter solution

Water Soluble Fertilizers

N:P:K + TE

19:19:19 + TE

Technical Aspects

Analysis

N	:	19 %
P_2O_5	:	19 %
K_2O	:	19 %
Trace Elements		
MgO	:	1 %
B	:	0.01 %
Mo	:	0.001 %
Cu	:	0.01 %
Mn	:	0.02 %
Zn	:	0.02 %
Fe	:	0.04 %

Agronomic Benefits

100 % water-soluble fertilizer in a crystalline form. The NPK is especially developed to meet with all the nutrient requirements of the crops, during the second stage after planting.

The metallic trace elements are in chelated form in which the trace elements are protected against fixation in the soil and remain available for the plants. It contains no chlorine or sodium and is most suitable for use in intensive crop production, and for the nutrition of saline-sensitive crops.

There are several differences between the fertilization of crops grown in greenhouses and those grown in open fields. The uptake of nutrients is higher in protected crops due to faster growth and higher yields. The same applies to crops grown in spring and summer, versus crops grown in winter. The nutrient requirements also depend on climate conditions, soil type, temperature, varieties, methods of application and irrigation.

Use :-

- Prepare a mother solution by adding 15 kg NPK 1:48:35 to 100L water (20°C). This solution then contains:
- 1.5 g. N/liter solution

- 72.0 g. P_2O_5/liter solution
- 52.5 g. K_2O/liter solution
- 1.5 g. MgO and a well-balanced micronutrient mix, developed to provide the crops with all the necessary trace elements in the right proportion.

Water Soluble Fertilizers

N:P:K + TE

20:10:10 + TE

Technical Aspects

Analysis

N	:	20 %
P_2O_5	:	10 %
K_2O	:	10 %
Trace Elements		
MgO	:	1 %
B	:	0.01 %
Mo	:	0.001 %
Cu	:	0.01 %
Mn	:	0.02 %
Zn	:	0.02 %
Fe	:	0.04 %

Agronomic Benefits

100 % water-soluble fertilizer in a crystalline form. The NPK is especially developed to meet with all the nutrient requirements of the crops, during the second stage after planting.

The metallic trace elements are in chelated form in which the trace elements are protected against fixation in the soil and remain available for the plants. It contains no chlorine or sodium and is most suitable for use in intensive crop production, and for the nutrition of saline-sensitive crops. As it contains no urea, the nitrogen in this NPK acts very rapidly.

There are several differences between the fertilization of crops grown in greenhouses and those grown in open fields. The uptake of nutrients is higher in protected crops due to faster growth and higher yields. The same applies to crops grown in spring and summer, versus crops grown in winter. The nutrient requirements also depend on climate conditions, soil type, temperature, varieties, methods of application and irrigation.

Use :-

- Prepare a mother solution by adding 15 kg NPK 20:10:10 + TE to 100 L water (20°C). This solution then contains:
- 37 g. N/liter solution
- 15 g. P_2O_5/liter solution
- 15 g. K_2O/liter solution
- 1.5 g. MgO and a well balanced micronutrient mix, developed to provide the crops with all the necessary trace elements in the right proportion.
- By applying 66, 7 1 of the mother solution, you supply the crops with 2.5 kg N, 1 kg P_2O5, 1 kg K_2O, 0.093 kg MgO and all the Trace Elements.

SWater Soluble Fertilizers

N:K

13 : 0 : 45

Potassium Nitrate

Technical Aspects

100 % macronutrient content

Potassium K_2O (available as K+)		45 %
Nitrogen N (available as NO_3)		13 %
Completely soluble in water with an insoluble Content		<0.075 %

Solubility at 20° C : 315 gram PK/liter of water

Free of Chlorine, Sodium and heavy metals

Chlorine (Cl^-)	max 0.5 %
Sodium (Na+)	max 0.5 %
Heavy metals	<10 ppm

Easy to mix, PK is compatible with all water soluble fertilizers

pH at 1 % solution in water	8-9

Low salt index and low EC value **E.C. value 1 g/1 at 25°C 1.3 mS/cm**

Potassium Nitrate's Key Advantages

High Purity Fertilizer :	**Potassium Nitrate contains no elements detrimental to plants: it is free of Chlorine** compounds, Sodium and heavy metals.
Fully soluble :	**Potassium Nitrate is a free-flowing, fine crystalline powder that dissolves quickly** in water.
Compatibility :	**Potassium Nitrate can be mixed with all water-soluble fertilizers, Potassium Nitrate** is compatible with the majority of pesticides in foliar application.
Nitrate Nitrogen :	**Nitrate Nitrogen is non-volatile and enhances the uptake of other cations (Ca^{++}, Mg^{++} ...).** Nitrate Nitrogen is the most efficient Nitrogen source for plant growth.
No Salt Accumulation :	**Potassium Nitrogen can be used to cover the potassium needs of a crop without supplying** excess of Sulfate or Chlorine.
All Round :	**Due to its low N/K ratio, Potassium Nitrate is suitable for all crops and growing stages, including** flowering and ripening stages.

Fertigation

Potassium Nitrate is an indispensable components in all well-balanced fertigation programs, both for soil-grown crops and hydroponics' systems. Potassium Nitrate, used as a K-source, will contribute part of the Nitrate Nitrogen. It is the only K-fertilizer suitable for intensive cropping without risk of Sulfate or Chlorine salt accumulation. Under saline or alkaline conditions, Potassium Nitrate is the only suitable K-choice.

For intensive horticulture, Nitrate Nitrogen is the optimum Nitrogen source. A high NO/NH_4 + balance assures sufficient uptake of Ca ++ and Mg ++ and avoids root degeneration, especially at high soil temperatures. Additionally, Nitrate Nitrogen losses by volatilization are avoided.

For hydroponics' systems, the most intensive and sensitive technique, this is even more important.

KNO_3 in a hydroponics' system is imperative and today's horticulture would not be the same without Potassium Nitrate.

Foliar Application

Due to low risk of leaf burning (low EC-value and osmotic pressure) and wide compatibility with other fertilizers and pesticides, Potassium Nitrate is a recommended foliar fertilizer.

Foliar application of Potassium Nitrate is an excellent way to supply additional Nitrogen and Potassium during critical physiological or environmental conditions.

Furthermore, important physiological processes such as breaking of dormancy (mango and deciduous fruits), ripening, fruit maturation, etc. can be influenced with foliar applications of Potassium Nitrate.

Water Soluble Fertilizers

NP

12:61:00

(Mono Ammonium Phosphate)

Technical Aspects

100 % macronutrient content

Nitrogen (NH_4)	12 %
Phosphorus (P_2O_5)	61 %
Completely soluble in water with a low insoluble content	< 0.01 %
Solubility at 20° C : 365 g/liter	
Free of Chlorine (Cl^-) and heavy metals	
Chlorine (Cl^-)	5-10 ppm
Heavy metals	< 10 ppm
Easy handling, Free flowing	0.05 %
Humidity	
pH at 1 % solution in water	4.5
Low salt index and low EC value	
Salt index	30 %
E.C. value 1 g/l at 25° C	0.86 mS/cm

MAP (Mono Ammonium Phosphate) is a fully water-soluble fertilizer, composed of 100 % plant nutrients with 84 % di-Hydrogen-Phosphate (H_2PO_4-) and 17 % Ammonia (NH_4+). MAP is therefore the water-soluble fertilizer with the highest P-content available. This, combined with its low Ammonia content, makes MAP suitable for fertigation in soil-grown crops.

Agronomic Benefits

Used mainly as a P-source, MAP is the perfect product to supply Phosphate and Nitrogen simultaneously in fertigation programs for soil-grown crops.

In intensive fertigation, control over N availability is limited if Urea is used as a Nitrogen source. MAP's low Ammonia content maintains the correct Nitrogen balance between Ammonia and Nitrate, avoiding the use of Urea. Combination with any Nitrate Nitrogen source ensures an optimum NO_3-/NH_4+ ratio for any crop.

MAP's Key Advantages

High Purity Fertilizer	:	**MAP contains no elements detrimental to plants: it is free of Chlorine compounds, Sodium and heavy metals.**
Fully soluble	:	**MAP is a free-flowing crystalline powder that dissolves quickly in water.**
Compatibility	:	**MAP can be mixed with all water – soluble fertilizers other than Calcium-containing fertilizers.**
pH Buffer	:	**The Phosphate in MAP buffers fertigation solutions, keeping the pH effectively stable at around 4.5.**
Urea-free	:	**Nitrogen in MAP is present in the Ammonia form, offering better control over N-availability to plants than Urea.**

Water Soluble Fertilizer
K
00:00:51 + 18 % S
(Potassium Sulphate)
Technical Aspects

100 % macronutrient content	
Potassium K_2O	51 %
Contains additional essential Sulfur ($S\text{-}SO_3$)	18 %
Solubility at 20° C : 111 gram K/liter.	
Free of Chlorine (Cl-), Sodium (Na+) and heavy metals	
Chlorine (Cl-)	1 % max.
Sodium (Na+)	1 % max.
Heavy metals	< 10 ppm
Easy handling, Free flowing	
Humidity	0.5 %
pH at 1 % solution in water	3.0
Low salt index and low EC value	
Salt index	30 %
E.C. value 1 g/l at 25° C	

Potassium Sulphate is a water-soluble fertilizer, composed of 100 % plant nutrients with 44 % Potassium (K_+) and 56 % Sulphate (SO_4-). Due to the absence of Nitrogen and the high Potassium content, Potassium Sulphate is a useful fertilizer for drip irrigation, especially in areas with sufficient water availability.

Agronomic Benefits

In fertigation, accurate control of the supply of every nutrient is a key to nutrition management. Potassium Sulphate is a Chlorine-free, water-soluble fertilizer... the only Potassium fertilizer that also provides the crop with Sulfur, an essential secondary macronutrient. As it is free of other macro-nutrients, Potassium Sulphate is a useful tool that allows the application of Potassium at the requested levels at any growing stage. Potassium Sulphate is an excellent fertilizer for providing part of the Potassium requirement, being careful not to exceed the Sulphate needs of the plant, thus avoiding excess.

It can be combined with any Nitrogen source according to crop requirement, controlling the K/N ratio as desired.

In drip irrigation systems on alkaline soils, the low pH of Potassium Sulphate contributes to a slight acidification of the zone surrounding the roots (rhizosphere), enhancing the availability of Phosphate and micronutrients present in the soil.

In hydroponics, Potassium Sulphate is used as a sulphate source for crops that demand high sulfur.

Potassium sulphate's key Advantages

High Purity Fertilizer	:	**Potassium Sulphate contains no elements Detrimental to plants: it is free of Chlorine Compounds, Sodium and heavy metals.**
Fully soluble	:	**Potassium Sulphate is a free-flowing, fine Crystalline powder, soluble in water.**
Compatibility	:	**Potassium Sulphate can be mixed with all water-soluble fertilizers, other than Calcium-Containing fertilizers.**
Flexibility	:	**The high concentration of Potassium and absence of Nitrogen and Phosphate enables the correct quantity of Potassium to be supplied at any stage. Potassium sulphate can be combined With N, P or NP- fertilizers according to crop needs**
Low pH	:	**In alkaline soils, Potassium Sulphate lowers the pH in the rhizosphere, helping to improved availability of Phosphates and certain trace elements. In drip irrigation solutions, it helps to reduce the pH of the irrigation water.**

NPK 19-19-19 + 1.5 MgO + TE

Specifications:

Nitrogen (N)	19.0
Ammonium, NH_4-N	3.7
Nitrate, NO_3-N	5.6
Urea, NH_2-N	9.7

Phosphorous (P_2O_5)	19.0
Potassium (K_2O)	19.0
Magnesium (MgO)	1.5
Sulphur (S)	1.2
Trace Elements:	
Iron (Fe) EDTA-chelated	0.04
Manganese (Mn) EDTA chelated	0.02
Zinc (Zn) EDTA chelated	0.02
Copper (Cu) EDTA chelated	0.01
Boron (B)	0.01
Molybdenum (Mo)	0.001
EC Value 1 gram/liter at 25°C (mS/cm)	

Water Soluble Fertilizers

P:K

0:52:34

Technical Aspects

100 % macronutrient content	
Phosphor P_2O_5 (available as $H_2PO_4^-$)	52 %
Potassium K_2O (available as K^+)	34 %
Completely soluble in water with a low insoluble content	< 0.1 %
Solubility at 20°C: 180 gram PK/liter.	
Free of Chlorine (Cl-), sodium (Na+) and heavy metals	
Chlorine (Cl-)	20-60 ppm
Sodium (Na+)	50-100 ppm
Heavy metals	< 10 ppm
Easy handling, Free flowing, no caking	
Humidity	0.2-0.5 %
pH at 1 % solution in water	4.5
Low salt index and low EC value	
Salt index	8.4 %

Agronomic Benefits

1. Increase of yield and better fruit quality. The efficient root uptake of both Phosphate and Potassium of PK in FERTIGATION guarantee an optimum metabolism and growth of plants and fruits (formation of nucleic acids, flowering, photosynthesis, respiration, formation and transport of sugars...).

2. The physico-chemical characteristics of 0:52:34, pH, high solubility and low salt index makes the product particularly suitable for:
   - Application as a foliar fertilizer (low risk for leaf burning and scorching)
   - Application in fertigation as a regular P source, especially for irrigation water with a high salt concentration since PK has a very low conductivity value (EC value).
3. PK has shown good effect in foliar sprays against powdery mildew in vineyards,

ORGANIC FARMING

Most countries have traditionally utilized various kinds of organic materials to maintain or improve the tilt, fertility, and productivity of their agricultural soils. However, several decades ago organic recycling practices in some countries were largely replaced with chemical fertilizers which were applied to high yielding cereal crops that responded best to a high level of fertility and adequate moisture, including irrigation. Soil tillage was also intensified to improve weed control and seedbed conditions. Consequently, the importance of organic matter to crop production received less emphasis, and its proper use in soi! management was sometimes neglected, or even forgotten.

Soil Organic Matter and Humus

Farmers since ancient times have recognized significant benefits of soil organic matter to crop productivity. These benefits have been the subject of controversy for centuries and some are still debated today. The following list includes many of the recognized benefits of soil organic matter.

1. It serves as a slow-release source of N, P, and S for plant nutrition and microbial growth.
2. It possesses considerable water-holding capacity, and thereby helps to maintain the water regime of the soil.
3. It acts as a buffer against changes in pH of the soil.
4. Its dark colour contributes to absorption of energy from the sun and heating of the soil.
5. It acts as a *cement* for holding clay and silt particles together, thus contributing to the crumb structure of the soil, and to resistance against soil erosion.
6. It binds micronutrient metal ions in the soil that otherwise might be leached out of surface soils.
7. Organic constituents in the humic substances may act as plant-growth stimulants.

TYPES OF BULKY ORGANIC MANURES

Farm yard manure (FYM), farm compost, town compost, night soil, sludge and green manures are bulky in nature and supply large quantities of organic matter but small quantities of plant nutrients in comparison to the inorganic fertilizers.

Farmyard Manure

Good-quality farmyard manure is perhaps the most valuable organic matter applied to a soil. It is the most commonly used organic manure in most countries of the world. It consists of a decomposed mixture of cattle dung, the bedding used in the stable and any remains of straw and plant stalks fed to cattle. It is one of the most important agricultural by-products. Unfortunately more than 50 per cent of the cattle dung produced in India is burnt.

Table- Average nutrient content of bulky manures

Manure	Percentage content		
	Nitrogen (N)	Phosphoric acid (P_2O_5)	Potash (K_2O)
Animal refuse			
Cattle dung, fresh	0.3-0.4	0.1-0.2	0.1-0.3
Horse dung, fresh	0.4-0.5	03.-0.4	0.3-0.4
Sheep dung, fresh	0.5-0.7	0.4-0.6	0.3-1.0
Night soil, fresh	1.0-1.6	0.8-1.2	0.2-0.6
Poultry manure, fresh	1.0-1.8	1.4-1.8	0.8-0.9
Raw sewage, fresh	2.0-3.0	-	-
Sewage sludge, dry	2.0-3.5	1.0-5.0	0.2-0.5
Sewage sludge, activated dry	4.0-7.0	2.1-4.2	0.5-0.7
Cattle urine	0.9-1.2	Trace	0.5-1.0
Horse urine	1.2-1.5	Trace	1.3-1.5
Human urine	0.6-1.0	0.1-0.2	0.2-0.3
Sheep urine	1.5-1.7	Trace	1.8-2.0
Wood ashes			
Ash, coal	0.73	0.45	0.53
Ash, household	0.5-1.9	1.6-4.2	2.3-12.0
Ash, wood	0.1-0.2	0.8.5.9	1.5-36
Farm, factory and habitation wastes			
Rural compost, dry	0.5-1.0	0.4-0.8	0.8-1.2
Urban compost, dry	0.7-2.0	0.9-3.0	1.0-2.0
Farmyard manure, dry	0.4-1.5	0.3-0.9	0.3-1.9
Filter-press cake	1.0-1.5	4.0-5.0	2.0-7.0

Plant residues			
Rice hulls	0.3-0.5	0.2-0.5	0.3-0.5
Groundnut husks	1.6-1.8	0.3-0.5	1.1-1.7
Straw and stalks			
Pearl millet	0.65	0.75	2.50
Banana, dry	0.61	0.12	1.00
Cotton	0.44	0.10	0.66
Sorghum	0.40	0.23	2.17
Maize	0.42	1.57	1.65
Paddy	0.36	0.08	0.71
Tobacco	1.12	0.84	0.80
Pigeonpea	1.10	0.58	1.28
Wheat	0.53	0.10	
Sugarcane	0.35	0.10	0.60
Tobacco dust	1.10	0.31	0.93
Tree leaves, dry			
Calotropis gigantea	0.35	0.12	0.36
Careya arborea	1.67	0.40	2.20
Cassia auriculata	0.98	0.12	0.67
Dollenea pentagyna	1.34	0.50	3.20
Madhuca indica	1.66	0.50	2.00
Pongamia pinnata	3.69	2.41	2.42
Pterocarpus marsuptum	1.97	0.40	2.90
Terimalia chebula	1.46	0.35	1.35
Terminalia paniculata	1.70	0.40	1.60
Terminalia tomentosa	1.39	0.40	1.80
Xylia dolabriformis	1.37	0.30	1.61
Green manures, fresh			
Cowpea (*Vigna catjang*)	0.71	0.15	0.58
Sesbania aculeata	0.62	-	-
Cluster-bean(*Cyamopsis tetragonoloba*)	0.34	-	-
Horse-gram (*Dolichos biflorus*)	0.33	-	-
Mothbean (*Vigna aconitifolia*)	0.80	-	-
Greengram (*Vigna radiata*)	0.72	0.18	0.53
Sunhemp (*Crotalaria juncea*)	0.75	0.12	0.51
Blackgram (*Vigna mungo*)	0.85	0.18	0.53

Composted Manure

Another method of augmenting the supplies of organic matter is the preparation of compost from farmhouse, and cattle-shed wastes of all types. Composting has been advocated and adopted extensively during the past 25 years. Composting is the process of reducing vegetable and animal refuse (rural or urban) to a quickly utilizable condition for improving and maintaining soil fertility. Studies conducted in India and elsewhere have shown that good organic manure similar in appearance and fertilizing value to cattle manure can be produced from waste materials of various kinds, such as cereal straws, cotton stalks, groundnut husk, farm weeds and grasses, leaves, leaf-mould, house refuse, wood ash, litter, urine-soaked earth from cattle-sheds and other similar substances. These materials are rich in cellulose and other readily decomposable carbohydrates and have a carbon-nitrogen ratio of 40 or more to 1. The direct application of such under composed, low-nitrogen organic matter as manure brings about a temporary deficiency of mineral nutrients (specially nitrates and ammonium compounds) in the soil by stimulating the growth of micro-organisms, which in turn, compete with crop plants for available nitrogen, phosphorus and other elements. Hence, before using them as manure, it is necessary to compost or partially decompose them. This process lowers the carbon-nitrogen ratio to about 10 or 12 to 1.

Urban Compost

In recent years, large-scale composting of town refuse and night-soil in properly constructed trenches away from human habitations has been taken up successfully by the municipal bodies of many large and small towns. Trenches, 1 to 1.2 m wide, 75 cm deep and of convenient length, are filled with successive layers of night-soil, town refuse and earth, in this order. The compost gets ready in about three months.

Sewage and Sludge

Domestic and industrial wastes (sewage and sludge) contain large quantities of plant nutrients and are used for growing of crops near many towns. In many places, the undiluted sludge has been found to be too strong for healthy plant growth and if it contains readily oxidizable organic matter, its use actually reduces nitrates present in the soil. The disadvantages are still greater if sewage is used on land without preliminary treatment. The soil quickly becomes sewage sick owing to the mechanical clogging by colloidal matter in the sewage and the developments of anaerobic organisms which not only reduce the nitrates already present in the soil but also produce alkalinity. Bacterial contamination makes the eating of raw vegetables grown on untreated sewage or sewage sludge a real danger to health.

Municipal and sewage wastes also form an important component of organic farming. The total and utilizable nutrient potential from garbage and sewage sludge works out to be substantial. The total municipal refuse is about 12 million tones/annum containing about 0.5 % N, 0.3 % K, whereas, sewage sludge amounting to 4 million tones/annum contains about 3 % N, 2 % P and 0.3 % K. Besides the necessity of sewage farming from nutrient utilization point of view, it is also an effective method to avoid pollution.

Night-soil

Very few towns in India are equipped with proper sewage system. The sanitary disposal of night-soil with an effective control of foul smell and fly nuisance is, therefore, a serious problem all over the country. Since human excreta is a potential source of soil improvement, public health authorities in several countries make the necessary arrangements for its conservation and conversion into a form in which it can be safely used as a manure. The dehydration of night-soil, as such, or after admixture with absorbing materials, e.g. soil, ash, charcoal, and sawdust, produces a poudrette that can be easily used as manure. The mixing of night-soil with an equal volume of ash and 10 per cent powdered charcoal produces and odourless material, containing 1.32 per cent nitrogen, 2.8 per cent phosphoric acid 4.1 per cent potash and 24.2 per cent lime. The addition of 40 to 50 per cent of sawdust to the night-soil yields straightway a dry, acidic poudrette which may contain 2 or 3 per cent nitrogen.

Green manuring

Green manuring can be defined as a practice of ploughing or turning into the soil un-decomposed green plant tissues for the purpose of improving physical structure as well as fertility of the soil. From the time immemorial the turning in a green crop for improvement of the conditions of the soil has been a popular farming practice. Green-manuring, wherever feasible, is the principal supplementary means of adding organic matter to the soil. It consists in the growing of a quick growing crop and ploughing it under to incorporate it into the soil. The green-manure crop supplies organic matter as well as additional nitrogen, particularly if it is a legume crop, which has the ability to fix nitrogen from the air with the help of its root-nodule bacteria. The green-manure crops also exercise a protective action against erosion and leaching. The adoption of green manuring depends upon the agroclimatic conditions. Broadly the following two types of green manuring can be thought of.

1. **Green manuring *in situ*:-** In this system, green manure crops are grown and buried in the same field which is to be green-manured, either as a pure crop or as an intercrop with the main crop. In India, various methods of growing green manure crops in situ are followed to suit local conditions. For the

proper decomposition of the green manure, it is necessary that the green material should be succulent and there should be adequate moisture in the soil. Plants at the flowering stage contain the greatest bulk of succulent organic matter with a low carbon/nitrogen ratio. The incorporation of the green-manure crop into the soil at this stage allows a quick liberation of nitrogen in the available form.

2. **Geen leaf manuring:** - Green-leaf manuring refers to turning into the soil green leaves and tender green twigs collected from shrubs and trees grown on bunds, waste lands and nearby forest areas. The common shrubs and trees used are Glyricidia (*Glyricidia maculata*), *Sesbania speciosa*, Karanj (*Pongamia pinnata*), etc. Green manuring in situ is followed in northern India while green leaf manuring is common in eastern and central India.

B. Concentrated Organic Manures:

Concentrated organic manures are those materials that are organic in nature and contain higher percentages of essential plant nutrients such as nitrogen, phosphorus and potash, as compared to bulky organic manures. These concentrated manures are made from raw materials of animal or plant origin. The concentrated organic manures commonly used are oilcakes, blood-meal, fish meal, meat meal and horn and hoof meal.

Oilcakes

These can be grouped into two classes, namely, (i) edible oilcakes-suitable for feeding the cattle and (ii) non-edible oilcakes- not suitable for feeding the cattle. Edible oilcakes are normally fed to cattle as concentrates. However, some of the edible oilcakes are also applied by the farmers to the soil. The quantity of nitrogen varies with the type of oilcake. It is from 2.5 per cent in mahua cake to 7.9 per cent in decorticated safflower cake. In addition to nitrogen, all oilcakes are the quick-acting organic manures. Though they are insoluble in water, their nitrogen becomes quickly available to the plants in about 7 to 10 days after application. The mahua cake, however is an exception, as its nitrogen does not become available till about two months after application. As such mahua cake should be applied about two months before sowing, provided the soil is moist. This cake is suitable for application to fruit plants and plantation crops.

How to Use Oilcakes: - These should be well powdered before application, so that they can be spread evenly and are easily decomposed by microorganisms. They can be applied a few days prior to sowing or at sowing time. Oilcakes-especially groundnut cake, are also applied extensively in the form of a top-dressing to sugarcane. Depending on crop, oilcakes are applied broadcast, drilled or placed while earthing up near root zone.

Cattle, Pig and Poultry Manures

In general, the higher the dry matter, the higher the nutrient content. For example, the higher nutrient content of the poultry manure is mainly due to the higher dry matter content. There is usually about the same level of nitrogen and potassium in cattle slurry, and phosphorus content is usually about 25 % of the nitrogen and potassium level. Pig manure usually has about the same level of nitrogen and potassium in cattle slurry, and phosphorus content is usually about 25 % of the nitrogen and potassium level.

Pig manure usually has about the same level of nitrogen as does cattle manure, although pig manure is usually higher in phosphorus and lowers in potassium than is cattle slurry. The difference in the phosphorus to potassium ratio between cattle and pig manure is principally reflection of the diet of these animals. The relatively low potassium and high phosphorus in pig slurry reflects mainly a cereal diet, and the high potassium in cattle manure reflects the high potassium content of herbage which often makes up a high proportion of the ruminant diet.

Meatmeal

An adult animal can provide 35 to 45 kg of meat after slaughter or death. At present, meat meal is manufactured in India on a small scale. The process of manufacturing meatmeal is very simple. First the bones and the meat are cooked or digested in special containers for two to three hours. The bones are then separated from the meat. This meat is dried on a sandbath till it is brittle and then it is powdered. The drying can be done in double-jacket trays worked by steam and the material dried over the steam for an hour. Mutton squeezer can also be used for removing water content of the meat, and then dried. Meatmeal is quick-acting manure and is effective for all crops, on all soil types. Its application is similar to oilcakes. Meatmeal contains about 10.5 per cent N and 2.5 per cent P_2O_5.

Bloodmeal

An adult cattle gives about 14.0 kg and a goat or sheep about 1.40 kg of blood. Slaughter-houses should be provided with pucca or concrete floor with a central drain leading into a blood storage tank. The blood is first treated with commercial copper sulphate @ 125 gm per 100 kg of blood. It is then evaporated to dryness on a sand bath. Next, it is spread on a concrete floor covered by a net, and allowed to dry in sun. When completely dried, it is powdered, bagged and marketed as bloodmeal. Bloodmeal is a quick-acting manure and is effective for all crops on all soil types. It should be applied like oil-cakes. Bloodmeal contains 10 to 12 per cent of N and 1 to 2 per cent of P_2O_5.

Fishmeal

Edible fish carcasses are used to prepare fishmeal. The fish is dried crushed or powdered and filled in bags. Fish manure is available either as dried fish, or as fishmeal, or powder. The manurial constituents present in it vary with the type of fish. Fishmeal is quick-acting organic manure and is suitable for application to all crops on all soils. It should preferably be powdered before use. Fishmeal contains 4 to 10 per cent, N, 3 to 9 per cent P_2O_5 and 0.3 to 1.5 per cent K_2O.

Horn-and-hoofmeal

A healthy animal gives about 3 to 4 kg of horn and hoof. These materials are cooked dried and powdered before bagging. This meal contains about 13 per cent N.

Sharma *et. al.*, (2009) studied the effect of organic manures and inorganic fertilizers on growth, physiological attributes, fruit yield and quality parameters of pomegranate Cv. Jalore seedless. The eleven treatments consisted of three organic sources i.e., sheep manure, cattle manure, vermicompost and inorganic fertilizer i.e., urea (N source) with two proportions i.e., 100% and 50:50 ratio and control (No nutrient source). The results revealed that highest plant height (2.55m) and canopy volume (18.50 m^2 were recorded in treatment containing vermicompost and inorganic fertilizers in 50:50 ratio. The maximum photosynthetic activity (0.65 mg CO_2 m^{-2} s^{-1}) water use efficiency (54.16%) and minimum transpiration rate (1.2 mg m^{-2} s^{-1}) was recorded in vermicompost and sheep manure in 50:50 ratio. The application of recommended dose of nutrient through vermicompost and inorganic fertilizer in 50:50 ratio gave the higher fruit yield (42.5 q ha^{-1}) over control and other treatments as well but statistically at par with vermicompost and sheep manure in 50:50 ratio. The application of nutrient through inorganic source preferably by vermicompost increased Tss (^{0} Brix), Juice content and reduced the percent juice acidity. Thus for better vegetative growth, physiological attributes fruit yield in pomegranate use of vermicompost and inorganic N in 50:50 ratio is suggested than their sole application in sandy soils of hot arid region.

Singh, (2009) studied the effect of organic nutrient sources on fruit yield and quality of pomegranate. The quality parameters of the fruits are genetically controlled by agricultural practices like use of fertilizers, manures, irrigation operations are important modifiers. It is better to use minimum quality of inorganic manures, pesticides and increase the application of organic manures and botanical pesticides to reduce the cost of production and to reduce microbial diversity, water contamination, chemical residue to reduce the health risk to human population.

Experiments were conducted for four years with following treatments. Vermicompost @ 18 kg/plant, Phosphocompost 21 kg/plant, Cattle dung manure @ 29 kg/plant, RDF (400:250:250 g) and 50% RDF + 50% CDM (INM) plant.

The doses of organic manures were calculated based on N equivalent basis; the yield and quality were assessed at harvest during winter season. Maximum yield was recorded in INM (7.98 kg/plant) treatment followed by organic and inorganic treatments which were on par with control and lowest (4.97 kg/plant). Fruit quality parameters such as total soluble sugars (Tss), ascorbic acid, and sugar content increased with application of organics where as anti-nutritional constituents such as juice acidity and tannin content adversely affect fruit quality. The maximum Tss were observed in cattle dung manure treatment (17.5 ^{0} Brix) and was lowest in control (1.55 ^{0}Brix). Higher accumulation of ascorbic acid was recorded in INM (17.9 mg/100 g) followed by organic and inorganic treatments and was the lowest in control (16.0 mg/100g). It concludes that application of 20 kg vermicompost or 20 kg phosphocompost or 30 kg cattle dung per plant or a combination of 50% CDM + 50% RDF improved the fruit yield and quality of pomegranate in vertisol.

Macro and Micronutrient Availability

Mineralization of plant nutrients and their availability to plants is one of the major consequences of organic recycling in soil. The quantity released depends mainly on the content of a particular nutrient in organic residues undergoing decomposition. In general, there is immobilization of nitrogen due to decomposition of cereal straw. However, with FYM/compost application no such effect is observed. The addition of FYM and cereal residues results in improvement of total soil nitrogen. The effect of organic matter in reducing the intensity of phosphate fixation by the soil sesquioxides and maintenance of soil fertility by use of organic manures along with super phosphate has also been established. FYM remains intermediate in building up available P status of soil. In sub-humid lateritic soils, P use efficiency under rice-rice sequence was increased tremendously with FYM application. Information need to be generated on what amount of fertilizer P could be curtailed for organic manures under different cropping sequences. Response of crops to organic manuring depends on time of application, soil characteristics, degree of decomposition of organic residues and C: N ratio and etc.

Continuous application of organic manures improves the availability of Zn in soil and may not be sufficient to meet the requirements of the immediate crops. Contrary to macronutrients, the range between deficiency and toxicity limits of micronutrients is quite narrow for most of the crops. Inclusion of micronutrients in fertilizer schedule should, therefore be advocated after careful appraisal through soil and plant tests.

Availability of organic waste in India

There are various estimates of agricultural waste availability. These may be based on crop residues, net availability on remote sensing techniques which is used in developed countries.

Estimates of agricultural waste availability suggest that the average value for crop wastes is 350 mt and that of animal wastes it is 650 mt. Hence, around 1000 mt of agricultural wastes are available in the country.

Agricultural waste availability in India

Agricultural Waste	Quantity (mt.)
Crop waste (including industrial waste)	301.6
Animal waste (including fisher is and marine waste)	944.4
Crop residues (including industrial waste)	108.0
Crop waste (cereals only)	289.7
Crop residues	407.0
Animal wastes	2018.0
Crop residues (including industrial waste)	347.2
Crop waste residues	287.3
Animal waste	633.0

Crop Waste : The quantity of wastes produced daily by an animal is a function of the type of animal, its size, its feed, and the temperature and humidity of the environment. Quantities quoted in the literature often refer to total weight or volume of fresh waste produced, including variable quantities of litter or water. Intensive livestock rearing tends towards standardization of diet and type of animal.

Cattle waste : The slurry production rate most often quoted for cattle is in the region of 40to 45 kg per day per livestock unit. A livestock unit (L.U) is equivalent to a 540 kg dairy cow. 500 kg beef cattle produces 40 kg of undiluted slurry per animal per day. It has been estimated that about 3 tonnes of dungstead manure or 6 tones of farmyard manure would be produced by one livestock unit during a 140 day winter period.

Pigs waste: The most often quoted slurry production for meal-fed pigs is 4.5 kg per pig per day on the average over the fattening period. At an average weight of 87 kg per pig. 4.35 kg slurry composed of 1.73 kg feces and 2.62 kg urine was produced. Pigs at 68 kg mean weight produced 4.5 kg slurry per day with 10 % dry matter. One sow and piglets produces about 14.3 kg slurry per day.

Poultry waste : Daily fresh poultry manure represents about 5 % of bird live weight. A 2kg mature bird would produce about 0.1 kg manure daily with 25 % dry matter. Broiler droppings are mixed with litter and normally contain about 75 % dry matter when removed from the house.

Silage Effluent: The quantity of silage effluent from ensiled grass and other materials is usually related to the dry matter of the ensiled materials; the lower the dry matter, the greater the quantity of effluent. A tone of ensiled material with 10 to 20 % dry matter will produce 260 to 450 kg of effluent. At 20 % dry matter, approximately 100 kg effluent per tone is produced.

Growing Mushrooms: Spent compost from growing mushrooms, along with other composts, are spread on land and can have a significant fertilizer value. The quantities are small by comparison with other agricultural wastes; however, they can be important in localized areas.

Utilization of Waste

- Chemical constituents like cellulose, hemi-cellulose and lignin lead to establish the potential agricultural wastes and food, feed, fodder, fiber and chemical products of economic importance. The calorific value (CV) indicates the fuel utility of agricultural waste. The CV for various agricultural waste materials varies between 14 to 19 MJ/kg. Higher C content favours calorific value of the waste material. The volatile water in plant residues may be around 80 per cent and fixed carbon is 20 per cent. Volatile matter in plant residues contribute towards oil and tar content. The characteristics of common agricultural wastes available in India.
- The characteristics of the agricultural waste materials make them important in respect of their use and economic value. Agricultural wastes can be put to use because of the following reasons:
- Farm development (improvement of soil health, soil fertility, soil-physical conditions a plant protection), it is the source of energy and power.
- It acts as animal feed and fodder.
- It is the human food.

Elemental composition of common agricultural wastes

Agricultural waste (biomass)	Nutrient Content (%)								
	C	H	N	Na	K	P	Ca	Mg	SiO_2
Pigeon pea stalks	53.30	4.70	0.60	0.05	0.57	0.08	0.11	0.41	0.68
Bagasse	48.20	6.10	0.20	0.06	0.51	0.04	0.14	0.36	1.30
Cotton sticks	51.00	4.90	1.00	0.09	0.61	0.08	0.12	0.43	1.33
Groundnut shell	41.10	4.80	1.60	0.05	1.20	0.12	0.10	0.40	2.52
Maize cobs	46.20	4.90	0.60	0.03	0.54	0.07	0.08	0.28	0.90
Maize stalks	41.10	4.20	0.60	0.04	0.42	0.05	0.09	0.45	2.00
Rice husk	37.80	5.00	0.30	0.02	0.30	0.03	0.08	0.17	15.60
Rice straw	36.80	5.00	0.59	0.09	2.50	0.06	0.10	0.53	16.77
Wheat straw	43.80	5.40	0.40	0.06	0.78	0.04	0.10	0.35	7.08

Some soil microorganisms play an important role in improving soil fertility and crop productivity due to their capability to fix atmospheric nitrogen, solubilise insoluble phosphate and decompose farm wastes resulting in the release of plant nutrient. The extent of benefit from these microorganisms depends upon their number and efficiency which, however, is governed by a large number of soil and environmental factors. When the number and activity of special microorganism called microbial inoculants or bio fertilizer is used to hasten biological activity to improve availability of plant nutrient. A number of products are now available that are generally referred to as soil and plant additives, of non-traditional nature. These products include (i) microbial fertilizers and soil inoculants which are purported to contain unique and beneficial strains of soil microorganisms, (ii) microbial activators that supposedly contain special chemical formulations for increasing the numbers and activity of beneficial microorganisms in soil, (iii) soil conditioners that claim to create favorable soil physical and chemical conditions which result in increased growth and yield of crops, and (iv) vermicompost which helps in improving soil health and fertility.

Nitrogen fixing organisms can be provided to the farmers in the name of microbial inoculants otherwise termed as Bio fertilizers. The bio fertilizers containing biological nitrogen fixing organisms are of utmost importance in agriculture in view of the following advantages.

1. They help in the establishment and growth of crop plants and trees.
2. They enhance biomass production and grain yields by 10-20 %.
3. They are useful in sustainable agriculture.
4. They are suitable in organic farming.
5. They play an important role in agro forestry/ silvipastural systems.

Biofertilizers

After the advent of the acetylene test for assessing nitrogen fixing potential, rapid progress has been made in identifying nitrogen fixing organisms. Out of a large number of microorganisms possessing the property of nitrogen fixation, only a few such as Rhizobium, Azotobactor, Azospirillum, BGA, Azolla, etc. have been commercially exploited. The reaction in biological nitrogen fixation is essential the same as in production of chemical fertilizers (Haber-Bosch process) i.e. the catalytic reduction of dinitrogen (N_2) to ammonia (NH_3).

Types of Biofertilisers

Rhizobium Inoculant

The most widely used biofertiliser is Rhizobium which colonizes the roots of specific legumes to form tumour like growths called root nodules. It is these nodules that act as factories of ammonia production. Rhizobia have the ability to fix atmospheric nitrogen in symbiotic association with legumes and certain non-legumes like Parasponia, Rhizobia colonize the roots of specific legume, enter through root hairs, multiply and form tumor like growth called root nodule. A mature nodule consists of a central bacteroid zone surrounded by several layers of cortical cells. The process of N-fixation is wholly dependent on the activity of the enzyme nitrogenous which is located within the bacteriods. The volume and number of bacteroids have a direct positive relationship with the amount of nitrogen fixed.

The Rhizobium-legume association can fix upto 100-300 kg N/ha in one crop season and in certain situations can leave behind substantial nitrogen for the following crop. The range of nitrogen fixed per hectare per year by different legumes is 100-150 kg for clover, 80-85 kg for cowpea, 100-300 kg for alfalfa, 240-320 kg for faba bean, 90-100 kg for lentil, 150-200 kg for lupins, 50-60 kg for groundnut, 60-80 kg for soybean, 50-55 kg for mungbean and 100-400 kg for pasture legumes.

Azotobacter Inoculant

Azotobacter is one of the most important non-symbiotic N-fixing micro-organism and considered to be very important for fixation of nitrogen in non-leguminous plants. Based on morphological and physiological features, the genus Azotobacter has been classified in 3 species i.e., Azotobacter chroococcum, A beijerinckii and A. vinelandii. The first two species are deemed to be the most commonly occurring species in India. Azotobacter chroococcum appeared in acid soils while A. beijerinckii in neutral and alkali soils. The beneficial effects of Azotobacter bio fertilizer on cereals, millets, vegetables, cotton and sugarcane under both irrigated and rainfed field conditions have been substantiated and documented. Application of azotobacter has been found to increase the yields of wheat, rice, maize, pearl millet and sorghum by 0.30 % over control. Apart from nitrogen, this organism is also capable of producing antibacterial and antifungal compounds, hormones and siderophores.

Azospirillum Incoculant

The ability to fix nitrogen by certain spirilla (*Spirillum lipoferum*) was first recorded by Beijerinck in 1925 and later on confirmed by several workers using not only conventional micro kjeldahal assay but also by the methods such as acetylene reduction technique and the isotopic enrichment method involving N. Tarrand and his associates (1978) renamed this organism as Azospirillum (N2 fixing Spirillum). Certain micro-organisms like bacteria and blue-green algae have the ability to use atmospheric nitrogen and transport this nutrient to the crop plants. Some of these nitrogen fixers like rhizobia are obligate symbiotic in leguminous plants, while others colonize the root zones and fix nitrogen in loose association with plants. A very important bacterium of the latter category is Azospirillum which was discovered by a Brazilian scientist and which attracted head-lines in the mid 1970s. The crops which respond to Azospirillum inoculation are maize, barley, oats, sorghum, pearl millet and forage crops. Azospirillum applications increase-grain productivity of cereals by 0-20 %, of millets by 30 % and of fodder by over 50 %.

Azolla-anabaena Symbiosis

Azolla is an aquatic fern commonly found floating in ponds, pools, tanks, shallow ditches and channels. It is also found in rice fields. Nitrogen fixing Blue-green Algae (GBA) *Anabaena azollae* is found in the cavities on its dorsal leaves and fixes atmospheric nitrogen efficiently which is available to Azolla plants. The recognition of Azolla contains 0.2-0.3 % nitrogen on fresh weight basis and 3-5 % nitrogen on dry weight basis.

A small floating water fern, Azolla is commonly seen in low land fields and in shallow fresh bodies. This fern harbours a blue-green algae, *Anabaena azollae*. The Azolla-Anabaena association is a live floating nitrogen factory using energy from photosynthesis to fix atmospheric nitrogen amounting to 100-150 kg N/ha/ year from about 40-60 tones of biomass. Reports of its current use as a bio fertilizer for rice in China, Vietnam, India, Indonesia, Thailand and other East and South Asian countries are available. An integrated system of rice Azolla fish has been developed in China.

Mycorrhizae : Special method for Obtaining Essential Elements

Mycorrhizae are the symbiotic association of fungi with roots of vasicular plants. The main advantage of mycorrhizae to the host plants lies in the extension of the penetration zone of the root fungus system in the soil, facilitating an increased phosphorus uptake. The interconnected net-work of external hyphae acts an additional catchments and absorbing surface in the soil beyond the depletion zone that would otherwise be inaccessible to the plant roots. Endotrophic mycorrhizae have been shown to be present in a wide range of horticultural species including apple, walnut, almonds, citrus, avocado, strawberry and grape. In many cases the mycorrhizae have been shown to markedly improve the growth of plants.

Phosphate Mobilising Biofertilisers

The efficiency of phosphatic fertilizers is very low (15-20 %) due to its fixation in soil. Besides, native soil phosphorus is mostly unavailable to crops because of its low solubility. The introduction of efficient P solubilizers in the rhizosphere has been found to increase the availability of phosphorus from both applied and native soil P. The phosphorus which gets fixed in soil is made available to crops by the action of microorganisms which solubilize insoluble/fixed forms of phosphates into forms which is readily taken up by the plants. Another factor which affects the effective utilization of P is its low mobility in soil. Fortunately some filamentous organisms like vesicular arbuscular mycorrhizae (VAM) which form symbiotic association with plant roots have the ability to mobilize phosphorus from soil and thereby helping in absorption of P by plant roots. Commercial exploitation of phosphates bio fertilizers can play an important role particularly in making the direct use of abundantly available low grade phosphate possible.

VERMI-COMPOST

It is a method of making compost with the use of earthworms which generally live in soil, eat bio-mass and excrete it in digested form. This compost is generally called vermin-compost or wormi-compost. It is estimated that 1800 worms which is an ideal population for one sq. meteoric feed on 80 tones of humus per year. To prepare an ideal vermin-compost, the following procedure is adopted).

Each shed measuring 20 ft x 80 ft is to be constructed with the help of locally available material like bamboos, stems of tress etc. A hut type structure is built with the help of these articles. The roof is made from dried grass, typha leaves, bamboo sticks, etc., in such a way that the hut may be protected from rain water and sun heat. Each hut may accommodate at least four vermin-beds of 3 ft. width.

These beds are prepared by putting 2 to 3 cm thick layer of farm manure as first layer followed by 10 to 15 cm of bio-mass with 200-250 worms per sq. ft. collected locally may be added and the bed should be kept sufficiently moist. This layer should be followed by a layer of 10 to 15 cm of half digested cow-dung layer which should be covered by a layer of leaves, trash etc. and water is sprinkled on the entire bed. The bed may be covered with palm leaves or coconut leaves or with any indigenous material. The pits should be kept constantly moist but never flooded.

A month later, the covered leaves should be removed and layers of organic waste not exceeding 6-7 cm should be added every alternate day. Watering should continue with each filling. When the pit is nearly full to a height of one meter, the material should be turned to provide aeration. After a month, the heap will be ready for harvest with good quality vermin-compost. The dug out vermin-compost should be heaped in an open place. The worms will find way to the bottom of the heap. The vermin compost from the top can be removed, dried and sieved for application in the field. The compost can also be enriched with micro-nutrients, bacteria etc. by adding them externally. 16 tones of compost can be obtained from 4 beds in 30 days after four months period of gestation.

Pulla Reddy *et. al.,* (2009) conducted experiment on effects of precursor – microbial interactions on induction of flowering in pomegranate. Precursor + microbial inoculum treatments recorded significantly superior results on various parameters observed precursor + inoculum significantly increased the leaf ethylene levels over precursors used alone. Application of methionine 2.5 ppm + inoculum and $FeSO_4$ 250 ppm + inoculum recorded significantly higher percentage of hermaphrodite flowers and intermediate flowers and reduced the male flower percentage. Highest percentage of flowering in pomegranate recorded in old wood by various treatments on new growth highest percentage of flowering and hermaphrodite flowers was recorded with methionine 2.5 ppm + inoculum and $FeSO_4$ 250 ppm + inoculum. Number of fruits per tree was highest in plants treated with methionine 2.5 ppm + inoculum followed by $FeSO_4$ 250 ppm + inoculum.

Biological methods

Bio pesticides take care of losses of crops, rejections in exports, losses of man hours and lives, losses of beneficial natural parasites and predators. Inspite of the thrust from the Govt. of India and State Government adoption of bio control agents and bio pesticides as an Agro input by farmers is still in infancy only about 5000 villages and one lakh farmers have reaped the knowledge and benefits of usage of Bio control agents and Bio pesticides. The real adoption of biopesticides by the farmers however will be evident only when it is used by them.

Nature has its own device to balance the ecosystem. Natural enemies like spiders, beneficial insects and parasites play a key role in maintaining the biotic balance and reducing levels of pest populations below the level of causing economic injury. Natural as well as applied biological control agents are important in pest and disease control. The natural enemies of pests viz., spiders and beneficial insects as predators, pathogens, parasitoids and the insect pathogens play a vital role in keeping the pest population at low level.

By way of encouraging spiders like long jawed spider (*Tetragnatha marillosa*), Orb spider (*Argiope sp*), Wolf spider (*Lycosa pseudoannualata*), Lynoc spider (*Oxyopes javanus*), Jumping spider (*Phidippus sp*), Dwarf spider (*Steypena sp*) and beneficial insects like Lady beetle (*Micraspsis sp*, *Harmonia sp* and Menochilus), Ground beetle (*Ophionea sp*), Mirid bug (*Ytrothinus sp*), Assasin bug (*Polytoxux tuscovitatus*), Ant (*Sofenopsis germinate*), Ear wing (*Euborellia spp.*) a good number can be controlled through out the season. These spiders and beneficial enemies are highly specific and are safe to non target species. Once established they survive in the nature till the pest is prevalent.

1) Fungal Bio-pesticides

a) *Trichoderma viride*

Trichoderma viride is available in wettable powder and liquid formulations.

Mycelian fragments, conidia and clenydospore of *Trichoderma viride* are highly active on root and stem rots caused by Schierotinia and Rhizoctonia. Wilts caused by Fusarium and Vesticillium and blights/leaf spots caused by Alternaria, Asochyta, Cercospora, Macrophonia, Myrothecium, Ramuleria, Downy mildew and Powdery mildews, Fungal diseases of cotton, cereals, pulses, vegetables, oilseeds, fruit plants and Floriculture. *Trichoderma viride* acts on the plant pathogens through competition for space and nutrients, parasitization, disintegrate the pathogen hyphae by enzymes and antagonism. Apart from this, Trichoderma viride metabolites stimulate seed germination, plant growth and early flower formation. Extremely safe to mammals, man, non-target organisms, pollinators, fishes and birds etc. *Trichoderma viride* is also classified as Class-IV, low hazardous (green label) pesticide. No phytoxicity and exempted from residue analysis. It can be safely used along with Bio-Fertilizers. *Trichoderma viride* is applied as paste or as slurry on seeds @ 4 ml/kg seed/seedlings as coating or as soil application @ 250 ml/ha. Do not use chemical fungicides before or after the use of *Trichoderma viride*. It should not be applied to the seeds treated with chemical fungicides.

Physical characteristics of liquid Trichoderma Formulation

- Very thin mate or no mate formation at surface of liquid formulation.
- Only spore appears in solution.
- Light green colour.
- More than 10^7 spores/ml with Nil contamination.
- Very high enzyme activity.
- Tolerates upto 70°C temperature.

b) Verticillium lecanii

Verticillium lecanii is available in wettable power and liquid formulation. Mycelial Fragments and spores of *verticillium lecanii* are highly active on sucking pests like scale insects, *Coccus viridis*, aphids, thrips, mites, bugs, jassids, hoppers etc., on fruit and vegetable crops. *Verticillium lecanii* infects all stages of insects including eggs, larvae, pupae, nymphs and adults. The spores attach to insect cuticle/integuments, germinate and penetrate the insect body. The hyphae through enzyme action, proliferate in the insect body and cause mortality through a combination of chemical, mechanical water loss and nutrient loss effects. The insect body ultimately gets covered with mycelium and spores, which may get dispersed and cause subsequent infections. Certain toxins (dipicolinic acid and bassianolide) produced by this fungus also cause the mortality of insects. Shelf life is 6 months, if stored in well-ventilated rooms is original packings at cool dry place at temperature below 40°C. Shelf life is one year for liquid formulations. It is extremely safe to mammals, human beings, natural parasites and predators, non-target insects, pollinators, fishes and birds etc. *Verticillium lecanii* is also classified as class IV, low hazardous (green label) pesticide. No phytotoxicity and is exempted from residue analysis. *Verticillium lecanii* is applied as sprays @ 5ml/L of water. Addition of jaggery or soap solution to the spray solution before spraying on the standing crop improves the results.

- Slightly reddish in appearance.
- No mycelium formation only spores in the solution.
- Releases much quantity of enzymes which ensures quick action.
- Adoptable to extreme environments.

2) Bacterial Bio-pesticides

a) *Bacillus subtilis*

The seed treatment with *Bacillus subtilis* stimulates plant growth to protect against soil borne pathogens has been a subject of research interest for over 20 years and as an inoculants for cotton, peanut and beans to suppress the Rhizoctonia and Fusarium rust disease complex. Applied as a seed treatment, the bacteria colonize the root system, competing with root-invading pathogens. The Bacterium produces resistant endospores tolerant to heat and desiccation and whose production is controlled by fermentation conditions. Spores are concentrated and dried in to a powder. The products may be applied as a water-based slurry or a hopper-box treatment in several alternative formulations at the rate of 10^4 – 10^5 spores per seed, and are compatible with a number of chemical seed treatments.

b) Pseudomonas fluorescens:

Pseudomonas fluorescens is available in wettable powder and in liquid formulations. Spores of *Pseudomonas fluorescens* are highly active on root and stem. Rots caused by Schlerotinia and Rhizoctonia, damping off caused by Pythium, blights/leaf spots caused by Alternaria, Ascochyta, Cercospora, Macrophonia, Myrothecium, Ramularia, Downy mildews and Powdery mildews, Fungal diseases of Cotton, Certals, Pulses, Vegetables, Oilseeds, Fruit plants and Floriculture. *Pesudomonas fluorescens* acts as the hyphae of plant pathogens by enzymes and antagonism. Seed treatment with pseudomonas fluorescens provides a protective zone around seeds. It also acts as a plant growth promoter and stimulates seed germination and early flowering and fruiting. Shelf life is 6 months, if stored in well ventilated rooms in original packings at cool dry place at temperature below 40°C. Shelf life is one year for liquid formulations. It is extremely safe to mammals, human beings, natural parasites and predators, non-target insects, pollinators, fishes and birds etc. *Verticillium lecanii* is also classified as class IV, low hazardous (green label) pesticide. No phytotoxicity and is exempted from residue analysis. It can be used only with Bio-fertilizers, pseudomonas fluorescens is applied as paste or slurry on seed @ 4 – 5 ml/kg seed/seedlings as coating or as soil application @ 250 ml/ha., after mixing with farmyard manure. It can also be used as foliar spray @ 200 ml/ ha. After the spray of pseudomonas fluorescens bactericidal chemicals should not be used and do not apply to the seeds treated with chemical bactericides.

Physical characteristics of liquid pseudomonas fluorescens

- Slightly greenish solution.
- When the formulation is mixed with water for use it release large amount of fluorescens which act quickly to control diseases.
- Nil contamination high count upto 10^{10}.

c) *Bacillus thuringiensis* var. Kurstaki.

Thuringiensis var. Kurstaki parasporal crystal containing Delta – endotoxin (b) spores high active ingredient content: potency 50000 1u/mg min. on *Helicoverpa armigera*, 55000 1u/mg min. on Spodoptera exigua 32000 1 u/mg min on Gypry moth). Highly active against $1^{st}/2^{nd}/3^{rd}$ instar Lepitopteran caterpillars including Helicoverpa, Spodoptera, Plutella, Borers, Hariy caterpillars, cut worms, Army worms, leaf-rollers and miners, skelotonizers and defoliaters, pest of cereals, pulses, vegetables, oil-crops, Fruit trees and Forest trees, *Bacillus thuringensis* var,Kurstaki acts on host larvae through its parasporal crystal delta endotoxin and bacterial spores causing starvation, paralysis and septicemia of larval bodies ultimately resulting in death of larva. The endotoxin acts on midgut epithelium causing disintegration of gut wall, cessation of feeding activity starvation and leakage of gut contents into haemolymph and body cavity of the larvae body. The spores germinate to form bacterial cells that multiply in the larval gut and body causing septicemia. Surviving larvae, if any, because of consuming sub-lethal doses, lead to under developed/deformed pupae/adults unable to form the progenies.

Shelf life is minimum 6 months, if stored in dark, well ventilated rooms in original packings at temperatures upto 40°C. For liquid it is more than 2 years at room temperature. Extremely safe to mammals, non natural parasite and predators (in non-target insects, fishes and birds etc. It is classified as class IV, low hazardous (green label) pesticide. No phytotoxicity and is exempted from residue analysis. Make homogenous thick paste of *Bacillus thuringensis* in small quantity of water. Dilute the paste in desired volume of water under intermittent stirring spray at the rate of 200 ml/ha. by knapsack or any other conventional sprayer at 12-15 days intervals based on pest populations. Spray preferably in evenings and on young larval stages or on sighting egg. *Bacillus thuringensis* var. Kurstaki results in instant stoppage of feeding and damage, death of caterpillars within 48-72 hrs of application. This result in decline of adult insects as well as decline in the crop damage.

Chapter 8

Irrigation and Methods of Irrigation

This crop can produce flowers and yields satisfactorily even under conditions of low soil moisture and saline soils (Guseva, 1960), but for optimum growth ample water is needed. Growth of plants was significantly reduced when soil moisture tension was greater than 2 atm. Observed that check basin required 108 ha. Cm of water, whereas drip with mulch required 40.31 ha. Cm. of water. Water saving was to the extent of 44 per cent in drip system and 65 per cent when sugarcane trash mulch was used in drip application of water equivalent to 20 per cent wetted area to be superior over surface method and average annual irrigation water requirement of pomegranate through drip method is 20 cm. Saving in water to an extent of 43 per cent and increased yields of 30-35 per cent is reported by them.

It is observed that more number of flowers and fruit yield under 0.8 IW/CPE ratio, which was increased to 1.0 during flowering and fruit development and decreased to 0.6 during bahar treatment. It is reported that in drip trial water use efficiency to be maximum (3.38 q/ha cm) in 20 per cent wetted area daily treatment followed by 20 per cent wetted area alternate day 93.32 q/ha cm).

Effect of saline water

High density in soils and irrigation with saline water affects normal fruit production. It is reported that irrigation with 6.5 mmhos for 8-10 months did not show any adverse effect on survival of cultivar Khog. Jalore Seedless could tolerate up to 4.5 mmhos electrical conductivity.

Fruit growth and development

The maximum growth rate (2.09 g/day) of the fruit is attained between 60-120 days stage in Ganesh and better fruit size can be attained by proper management during this period of active growth.

Water/moisture conserving measures

Black polythene followed by sawdust and banana trash proved to be the better mulch material (Chattopadhyay *et al.*, 1992). Ten per cent kaolin, 1.5 per cent power oil and 1 per cent liquid paraffin were found beneficial to increase productivity (Anon., 1995).

METHODS OF IRRIGATION APPLICATION

i) Surface irrigation: Water is applied by complete wetting land surface as in border strip or check method or wetting only a part of the surface as in furrows or ring.

ii) Sprinkler or overhead irrigation: The soil is wetted in a similar fashion as by rainfall.

iii) Sub-irrigation: Water is applied beneath the land surface from shallow water table wetting the surface little or not at all e.g. perforated pipe below soil to maintain water table.

Water management studies in important fruit crops

Sl. No.	Crops	Irrigation scheduling approach	Layout	No. of irrigations	Water requirement (cm)	Remarks
1	Pomegranate	50% soil moisture depletion.	Check basin	30	200-225	Uneven interval of irrigation during later period of fruit, depletion development and maturity induces cracking of fruits. Interval of 7-10 days in Mrig bahar and 5-7 days during Ambe bahar crop is recommended.

MATURITY INDICES

For getting quality fruits, they are harvested at proper stage of maturity. In Maharashtra, the maturity of pomegranate fruit is judged by the following indices:

1. The fruit requires 130-150 days after fruit set for proper maturity.
2. In summer the fruit colour changes from yellowish to dark red at maturity.
3. The fruit is harvested when it gives cracking sound when tapped.
4. The calyx at the distal end of the fruit gets closed at the time of maturity .
5. Properly matured fruit is easily scratched with finger nails.

Effect of method of irrigation on growth and yield of pomegranate

Methods	Litres water used (5 Mar.-14 June)	Tree height (cm)	Stem dia (cm)	Fruit yield (q/ha)	W.U.E. (q/ha/cm)
Check basin					
0.6 IW/CPE	48900	2.35	3.78	40.6	0.82
0.8 IW/CPE	6600	2.56	4.02	47.5	0.69
Trickle					
Daily, 20 %wetted area	3580	2.16	3.79	52.5	1.25
Daily, 30% wetted area	5322	2.44	4.02	44.9	0.72
Alternate day, 20% wetted area	359	2.30	3.84	48.6	1.16
Alternate day, 30% wetted area	5322	2.24	4.01	57.3	0.92

Performance of drip method of irrigation on different fruit crops

Sl. No.	Crops	Yield (q/ha)		Water applied (cm)		Water saving (%)	Increase in yield (%)
		Surface	Drip	Surface	Drip		
1	Pomegranate	53.5	69.7	218	121	45	12

Average water requirement of fruit tree by drip method (litre/tree/day)

Fruit Crops (spacing)	Age (yrs)	Water requirement (litres/tree/day)											
		Jan.	Feb.	Mar.	Apr.	May	June	July	Aug	Sept.	Oct	Nov.	Dec.
Pomegranate ($5x5m^2$)	1	-	-	-	-	-	-	13	10	10	12	9	8
	2	12	17	25	34	36	25	18	15	15	18	13	11
	3	16	23	34	46	50	34	25	21	21	24	19	16

CASE STUDY FOR POMEGRANATE ORCHARD

Design drip system of irrigation for 1 ha of an orchard planted with pomegranate (G-137). The spacing for pomegranate is 5 x 5 m. The available data is given below:

1. Size of land : 100 x 100 m

2. Type of soil : Light textured
3. Land slope : 0.30 per cent (S-N)
4. Maximum evaporation : 12.5 mm/day
5. Water source : Well
6. Available discharge : 3.5 lps
7. Static head : 10 m
8. Wetted area : 20 per cent
9. Age of orchard : 3 years

Net depth of water

i.e. Evapo-transpiration of crop (ET)

$E_T = E_p \times K_p \times K_c$

Where,

E_p = Pan evaporation (mm)

K_p = Pan factor (taken as 0.7)

K_c = Crop factor (depends on the age of the crop)

ET = 12.5 x 0.7 x 0.75

= 6.56 mm/day

The crop factors K_c and wetted area for different fruit crops are given as follows.

Sl No.	Fruit crops	First year		Second year		Third year		Fourth year		Fifth year on Wards	
		KC	Wa	KC	Wa	KC	Wa	KC	Wa	KC	Wa
1	Pomegranate ($5 \times 5\ m^2$)	0.40	0.30	0.50	0.35	0.60-0.65 (Full grown)	0.40	-	-	-	-

Volume of Water

Total volume of water required (mm) = Depth of water x Plant spacing (m) x Row spacing (m) x Wetted area (in fraction)

= 6.56 x 5 x 0.20

= 32.81 /day/ plant

Considering 90 per cent emission uniformity of the drip system, volume of water required = 36.51/ day/plant. Per plant wetted area is the area which is shaded due to its canopy cover, when the sun is overhead, which depends on the stage of the crop growth.

Emitter selection and positioning:

The age of pomegranate tree is 3 years i.e. the trees are fully grown. The soil is light textured.

Use 4 drippers, each of 4 lph dischanrge i.e. the total discharge rate = 16 1/h

Operation time of system

$$\text{Operation time} = \frac{\text{Volume of the water to be applied per tree}}{\text{Total dripper discharge } (4 \times 4)}$$

$$= \frac{36.5}{16}$$

$$= 2.28 \text{ h} = 2 \text{ h } 17 \text{ min.}$$

Drip Irrigation

The amount and distribution of rainfall in many parts of the world are inadequate to meet the total water requirements of crops. The only solution to this is by supplementing water through irrigation. Being a limited resource, efficient use of water through scientific irrigation management is of utmost importance in providing the best insurance against weather induced fluctuation in food production.

Drip or Trickle irrigation is one of the latest innovative irrigation methods. It is the slow application of water on or beneath the discrete or continuous drops, tiny streams or miniature spray through emitters or applications placed along a water deliver line near the plants. In this system water is delivered to each plant at it's root zone through a network of tubing.

PRINCIPLES OF DRIP IRRIGATION

Water is required only for the root zone of any crop and this can be achieved through drip irrigation system. It is a method of watering plants frequently and with a volume of water approaching consumptive use of the plants, thereby minimizing such conventional losses as deep percolation, run off and soil water evaporation. In this method irrigation is accomplished by using small diameter plastic lateral lines with devices called 'emitters' or 'drippers' at selected spacing to deliver water to the soil near the base of the plants. The system applies water slowly to keep the soil moisture within the desired range for plant growth. Drip irrigation system facilitates in maintaining a constant moisture level at the crop root zone thereby making available a regular supply of water and plant nutrients throughout the growth making available a regular supply of water and plant nutrients throughout the growing plant of the crop. Drip irrigation method is characterized by the following features:

i) Water is applied at a low rate
ii) Water is applied over a long period of time
iii) Water is a applied at frequent intervals
iv) Water is applied near or into the plant root zone
v) What is applied by a low pressure delivery system (0.2 to 2 kg/cm^2). Since the area wetted by each emitter is a function of the soil hydraulic properties, one or more emission points per plant may be necessary.

ADVANTAGES OF DRIP IRRIGATION

The potential advantages of drip irrigaition can be summarized as follows:

- Increased beneficial use of available water by irrigating only a smaller portion of the soil volume, decreasing surface evaporation and controlling on deep percolation losses.
- Enhanced plant growth and yield by keeping soil water content fairly constant in the plant root zone.
- Retarded weed growth since only a fraction of soil is wetted.
- Improved application of fertlizers by factilitating frequent split application with the irrigation water or otherwise.
- Decreased energy requirements since the quantity of water pumped is lesser than other irrigation methods and also due to the low operating pressure to drip irrigation system.
- Reduced operational labour when compared to conventional irrigation methods.
- Suitable for irrigation of hilly terrain and problem soils: If you take Kerela into consideration where high value plantation crops like coconut, arecanut, rubber, cardamom etc. are cultivated on a large-scale mainly on lands with undulating topography, the feasibility of adopting conventional methods of surface irrigation in these areas is less. In addition, the frequent occurrence of erratic monsoons and the existence of a definite dry spell during December-May season with problems of water scarcity necessitates economy in water use in this region. Drip irrigation is an ideal technique under such conditions.

Organo-mineral pits-A technique for soil moisture conservation

For better moisture conservation as well as application of irrigation water at root zone depth, under drip irrigation, water is applied in pits taken around the plant. In the case of coconut where root zone extends laterally to about 2.0 m radius, four pits are dug, each one at a distance of 1.5 m., from the palm on four sides of the basin. The size of each pit will be 0.5 x 0.56 x 0.5 m. These pits are filled with alternate layers of coconut husk, dry cow dung dry leaves and finally soil on top. These are called organo-mineral pits. Irrigation water is applied through the emitters into these pits. In order to enable the distribution of water to the lower portion of the pit s a sub-soil distributor is used.

Sub-soil distributor. This is a PVC tube (Half inch internal dia) of 25-30 cm length with perforations made all around. This tube is inserted into the organo-mineral pit and emitter is placed into it. This will enable better distribution of water to the lower portion of the pits and at the active root zone depth.

SYSTEM LAYOUT

Layout of the drip irrigation system involves simple pipe arrangements and the decision as to the number and spacing of laterals is governed by the nature of the crop. Each row of plants, whether trees, vegetables or others will usually require one lateral. In some horticultural crops, double row planting allows two crop rows to be served by the lateral. The in-row placement of emitters has the distinct advantage of minimum interference with other cultural operations. If adequate wetting is not achieved by one lateral line, emitters may be installed on a double line or a loop (some-times called a pig-tail) around the base of the tree.

FERTILISER APPLICATION UNDER DRIP IRRIGATION

Soluble fertilizers may be distributed through drip irrigation system, while those that are only partially soluble should not be applied through the drip system. With the commerical drip irrigation systems, available in the market, fertilizer injection tanks are provided, whereby the soluble fertilizers can be injected into the irrigation water. The partially soluble/insoluble fertilizers may be applied manually inside the organo-mineral pits where the drip emitters are placed.

Serra Hepaksoy *et al* (2006) studied the effect of irrigation levels on physiological responses in pomegranate varieties in Turkey. Pomegranate a tropical fruit adopted to Mediterranean climate and it shows good reponse to irrigation and stand drought tolerance. Irregular irrigation interval results in fruit cracking. First irrigation reached field capacity (S100) while the second reached the 40% (S40) and third non application (S0). Applications were made between June beginning and end of September irrigations were made at 20 days intervals during the irrigation period. They determine the response of the pomegranate varieties to the irrigation levels photosynthesis (A - $\mu mol^{-2}\ S^{-1}$) and transpiration (E – $mmols^{-1}\ S^{-1}$) (C1 D, portable photosynthesis system model (1-301) and leaf water potential (PMS instrument Inc Model 1003) were monitored during irrigation period. Leaf succulence and water deficiency in leaf tissue (%) stomata per unit area were determined.

Warade *et al.*, (2009) studied the effect of different mulches on yield and quality of pomegranate. They revealed mulching play predominant role in conserving moisture reducing weeds and controlling nematodes in pomegranate. Deficiency of moisture in the soil during fruit development stage is responsible for fruit cracking and reduces the fruit quality. Besides, controlling the weed population, mulching with black polyethylene paper reduces nematodes population which is the serious cause of wilting in pomegranate. By considering this aspect, four different types of mulches were used in the pomegranate orchard. The treatment differences were significant in various aspects. Mulching with black polyethylene sheet was found superior in respect of growth of plant and yield of fruits. Significantly highest marketable and total yield 15.55 and 17.09 kg/tree, respectively was recorded by the treatment of mulching with black polyethylene paper and was at par with treatment of mulching with Sugarcane trash, Soyabean straw and Wheat straw. The average number of fruits (58.75) and average weight of fruits (268.75 g) was also found maximum in treatment of black polyethylene mulch. The intensity of nematode population was maximum when crop was mulched with black polyethylene sheet.

Meshram *et al.,* (2009) carried out a study on water requirement of pomegranate (*Punica granatum L*) to select a best reference crop evapotranspiration model (Et_r), developed crop coefficient (k_c) and actual evapotranspiration of pomegranate (Et_c) for Solapur (North latitude 17°10″ to 18°32″and East Longitude 74°42″ to 76°15″) at 483.5 m in a msl in semi arid zone of Maharashtra. Six most commonly used reference crop evapotranspiration models were selected testing their validity under the climatic conditions. The important reference evapotranspiration models are: (i)Penam – Montheith FAO – 56 model (ii) modified penam FAO model (iii) Hargreaves samani model (iv) FAO – 24 pan evaporation model (v) Blancy criddle model (vi) FAO Radiation model. For this estimation of shaded area at solar noon hour, a plywood board of 3.5 x 3.5 m size with grid marking of size 10 x 10 cm was used. Actual pomegranate evapotranspiration was determined by the crop coefficient approach where the effects of various weather conditions were incorporated into reference crop evapotranspiration and crop characteristics into crop coefficient. Testing of these models were made on the basis of least root mean square error and regression analysis. R radiation method are higher followed by modified penam method. Blancy criddle, pan evaporation, Hargreaves – samani and Penman Moneteith. Out of six models Hargreaves samani method were found least root mean square error under climatic condition of Solapur region. The measured shaded area at solar noon increased from new leaf initiation to maturity of crop ranging from 0.68 to 10.8 m^2 and decreased from harvesting period 10.80 to 7.43 m^2. The water requirement of crop development and maturity period ranges from 32.85 to 65.75 liters/day/tree. The crop coefficient values increased from 0.15 to 1.20 from new leaf initiation to maturity period of crop and seasonal water requirement for pomegranate crop during kharif season varying from 9.6 to 65.75 liters/day/tree. The present study is based on analysis on long term (25 years) climatic data and measured shaded area at solar noon of 5th year pomegranate orchards.

Patel *et al.*, (2009) studied irrigation management through different methods of irrigation on pomegranate Cv. Ganesh in Gujarat, India. Water requirement in pomegranate through drip and surface methods of irrigation. Drip irrigation at 0.3, 0.4, 0.5, 0.6, and 0.7 PEF was listed along with surface method irrigation and no irrigation (rainfed). Surface method of irrigation gave significantly higher yield amongst all the treatments but it was at par with 0.7 drip ratio. There was no significant difference between 0.7 PEF and surface method of irrigation. Number of fruits per tree and average fruit weight were significantly highest through surface method of irrigation, but statistically at par with 0.7 PEF. Irrigation with 0.7 PEF could result in about 49% saving of irrigation water. Thus under water scarcity condition, irrigation through drip at 0.7 PEF can double the area under irrigation for pomegranate.

Hiwale (2009) studied root distribution pattern in pomegranate Cv. Ganesh. Root distribution act as guide for application of fertilizers as well as irrigation. Root distributions vary according to the type of soil, method of propagation. No such work was carried out in India on pomegranate. Root distribution in air layered of pomegranate Cv. Ganesh under semi-arid rainfed condition revealed that the root system is shallow in nature as below 60 cm soil depth not much root activity was recorded. Maximum root activity on fresh and dry weight basis was observed in 30 cm radial distance from tree trunk (54.17%) which was 31.12% in 30 – 60 cm radial distance and 2.12 % in 60 – 90 cm radial distances. However soil depth wise it was maximum at 30 – 60 cm and 42.69% at 0 – 30 cm depth it was least (3.74%) at 60 – 90 cm depth.

Chapter 9
Pomegranate Physiology

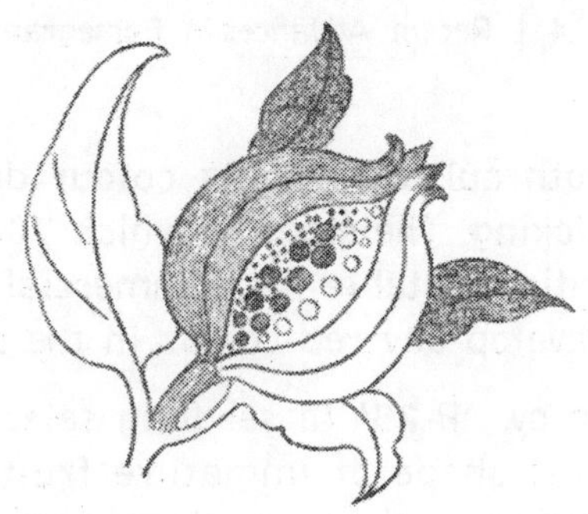

The fruits are ready for harvest between 135 and 170 days after anthesis. Fruits are harvested when the skin turns slightly yellow and the fruits give a metallic sound when tapped. At maturity the fruit colour changes in summer to dark yellow and in winter to dark red. The buds at the anterior end of the fruit curve inwards and become hard and dry at maturity. Properly matured fruits are easily scratched with a finger nail. The fruit is generally clipped from the tree when ripe and its quality improves on storing. Early harvesting in order to avoid cracking is one of the causes of poor quality of pomegranate in India.

Depending on the 'Bahar' treatments, pomegranate comes to harvest in installments. Usually the harvest commences in December/January and extends up to June/July. Crops ripening from April to June often get sunburn and may also crack if rains intervene or irrigation is irregular. The pomegranate cv. 'Wonderful' fruit reaches horticultural maturity for commercial harvest when the total soluble solids (TSS) content attains a fairly constant level of 15%. The level of titrable acidity varied from one location to another and from one year to the next but generally remains stable at the time when the T.S.S. content reached 15%. After harvest, there was no further change in either TSS content or titrable acidity at 20°C, but redness of the juice continues to increase upto the harvest.

The development of pomegranate fruit under continental (Bet Shean Valley) and moderate maritime (coastal plain) climatic conditions. They observed that fertile flowers were vase-shaped and developed fruits, while bell-shaped flowers contained few egg cells and were sterile. Gibberellic acid induced the sterile flowers to develop into small fruits; however, these were devoid of juicy seeds. The growth of 'Mule's Head' pomegranate followed a single sigmoid curve, whereas in cv. 'Wonderful' the growth was more linear. The seeds accounted for about half of the fruit weight. The edible juicy tissue of the seeds grew continuously from June to October, whereas the internal stone tissue ceased growing and hardened by the end of June, Juice, TSS and anthocyanin content of the fruit increased continuously during maturation, while acidity devised. Fruits of cv. 'Mule's head' ripen early and have a low acid content, whereas fruits of the late ripening cv. 'Wonderful' have high juice, TSS, acids and anthocyanins and are therefore suitable for processing. In both cultivars fruits colour developed gradually and served as a criterion for picking. The stage at which 70-90% of the skin is red corresponds with TSS: acid ration suitable for commercial picking. Some cultivars, such as 'Malissi', do not develop any red colour in the skin however.

In cv. 'P-23' (a seedling selection from the local Muscat') the initial elongated oval shape of immature fruit changed at maturity to spherical, with marked depressions on the sides, and the green-purple rind colour turned to deep pink with reddish and yellowish patches. The early milky-white colour of arils turned to a creamy pearl-white when the fruits matured. With the advancement of maturity, the percentage of arils increased while that of seed, rind and rind thickness decreased gradually. The TSS: acid ratio increased from 14.96 at 30 days to 64.83 at 165 days after anthesis.

From the pattern of CO_2 and ethylene production rates, pomegranate, fruits are judged to be non-climacteric fruits and thus maturation and ripening should take place on the plant before harvest to get quality fruits. The mature pomegranate may remain on the tree for a long period and picking time can be selected according to commercial needs. The harvesting of pomegranates in California begins in the middle of September and harvest maturity is judged on the basis of juice colour (the red colour must be equal to, or darker than, Munsell colour—chart 5 R-5/12), titrable acid content (<1.85%) of the fruit.

The red colour of pomegranate peel and juice is due to presence of anthocyanins. The most common anthocyanin in the juice is delphinidin-3, 5-glucoside and delphinidin-3-glucoside were also noted. The fruit rind contained large amounts of excessively astringent tannin, which made the juice unpalatable if the whole fruit is crushed or pressed at high pressure. The fruits 45-61% juice is calculated on a whole-fruit basis, or 76-85.5% in relation to the weight of arils. The juice has 12-16% sugar, consisting mainly of glucose or fructose; citric and malic acids have been identified as the predominant acids in fruits.

Post harvest storage

Pomegranate fruits are susceptible to moisture loss and need to be stored at high humidity. After harvest, the fruits are graded according to size, wrapped in paper and packed in bamboo baskets or corrugated boxes. In bulk storage, fruits are packed in layers in wooden crates, each containing about 16-18 kg of fruits. Dry grass, rice straw or papers are used as cushioning material. In the state of Maharashtra (India) pomegranates are sorted into four grades, viz, super, king, queen and prince. The fruits are then packed in corrugated fiberboard (CFB) boxes. Four of five super-size fruits, six king size, nine queen-size or twelve prince-size fruits will generally fill a box. Innovative packaging developed at the Indian Agricultural Research Institute using ventilated CFB boxes with CFB partitions having vent holes and layer separators could be very useful for packaging pomegranates.

Pomegranate fruits have better keeping qualities than other tropical fruits such as mangoes, grapes and bananas. The storage life of pomegranates is comparable to that of apples. They can be stored for some months in a cool dry place and upto six months under cold storage. The fruits, however, may be spoiled during transit or before marketing by rotting or the development of black colour in the fruits. The latter is thought to be due to post harvest biochemical changes.

Pomegranates can best be stored at low temperature and high humidity. Fruits stored at 4.5° C and 80-85% RH did not undergo any shrinkage or spoilage in a few months. It recommends an RH of 85-90% for storing 'Kandhar' pomegranates. Storage at lower temperature results in chilling injury, characterized by discolouration and pitting of the rind internal browing of the pith, paleness of the flesh and increased susceptibility to decay. Storage at 10°C is satisfactory if a post-harvest fungicide is used. Packing material containing sulphur compounds are found to give fruits 50% more protection from pathogens.

The control of RH is critical in the storage of pomegranate fruits. At low humidity, the skin dessicates readily and the rind becomes dark and hard; the fruits become less attractive and poor marketability. Storage at 5°C or lower resulted in chilling injury to the fruits. The severity of the symptoms increases with the exposure time and at temperatures below 10°C. The peel of 'Wonderful' pomegranate fruit undergoes browning during storage below 10°C.

The storage life of Banling' pomegranate could be extended up to 12 weeks by keeping in sealed polythene bags at 10°C, with slight changes in quality such as weight loss, TSS, titrable acidity. Individual shrink-warping of fruits and storing at 9°C gave 70 days of shelf-life without affecting the quality. Fruits dipped in 4% $CaCl_2$ solution for 3 minutes and stored at 2-4°C and 85-90% RH for 110 days showed low storage loss and retained quality. Fruits dried at 10°C, or 10°C and 47% RH, remained acceptable for three months or more during storage depending on the cultivar. Drying affects the quality but partially dried pomegranate may be useful for processed products.

Post harvest Physiology

The respiration rate of 'Wonderful' pomegranates remains low (less than 8 ml CO_2 $kg^{-1}h^{-1}$) when stored at 0-10°C for three months; it increases with temperature, however. The Q_{10} values for respiration are 3.4 between 0 and 10°C, 3.0 between 10 and 10°C and 2.3 between 20 and 10°C The rate of ethylene production from fruits stored below 10°C is mostly below 0.1 up $kg^{-1}h^{-1}$. Chilling injury symptoms increased susceptibility to decay organism become more apparent after transfer to 10°C, internal symptoms are manifested as pale colour of the arils. Fruits held at 5°C for eight weeks show only a slight brown discolouration of the loculars septa. The temperature during storage for up to three months has little effect on soluble solid content, pH and titrable acidity of the juice.

Physiological Disorders

When the physiological disorder known as 'internal breakdown' occurs in the pomegranate, the pulp-bearing seeds (arils) do not develop the typical red colour and are somewhat flattened rather that plump. Flavour of the arils is abnormal and many have a streaked appearance due to fine white lines radiating from the seeds. There are no external symptoms. It originates during growth in some seasons, usually only in limited areas.

The incidence of internal breakdown develops 150 days after anthesis in variety G-137, and its intensity increases in the fruits which are left on the tree upto 165 days. The incidence of browning increases with increase in weight of fruit from 150 to 200 g (26.60 %) to more than 350 g (60%) TSS, acidity, ascorbic acids, total sugars, reducing sugars, calcium, phosphorus and the enzyme catalyses were low whereas non-reducing sugars, starch, tannins, nitrogen, potassium, magnesium, boron, polyphenoloxidase and peroxides enzymes were high in affected arils of cvs. 'Ganesh' and 'P-23' than in healthy ones.

A superficial browning disorder (scald) develops on the husk of 'Wonderful' pomegranate fruit during storage. The severity of this disorder can be diminished by delaying the harvest time and by reducing storage temperature but these measures were insufficiently effective for storage periods exceeding six weeks and at temperatures of 9°C or lower chilling injury also occurs. Scald incidence was correlated with the amount of 0-dihydroxyphenols extractable from the husk, and was significantly controlled by measures that inhibited their oxidation by poly phenoloxidase. Such post harvest measures included dipping the fruit in boiling water for 2 minutes and in anti-oxidant solutions in biodithiocarbamate for 30 seconds, or by storing the fruit in a low-oxygen atmosphere. The most effective control of husk scald was obtained by storing late-harvested fruit in 2% O_2 at 2°C, but this treatment resulted in accumulation of ethanol, which caused off-flavors. When the fruits were transferred to air at 10°C both ethanol and off-flavours dissipated.

Thioproline (TCA) plays an important role in overcoming environmental stress by interacting with Proline metabolism and leading to an increase of its level in the plant. L-Cysteine activates various physiological and biochemical processes of plant cell resulting in increase of functional molecules like Aspartic and Glutamic Acid, Indol Acetic Acid (IAA) and Alpha Napthyl Acetic Acid (NAA), which ultimately lead to increased productivity and better quality of the final produce.

Functions of these amino acids.

Proline – FANTAC increases the secretion of Proline, which increases plant resistance to water and thermal stress.

Glutamic and Aspartic Acid – FANTAC increases Glutamic and Aspartic Acid content in plant cell which allow a better reserve of cellular Nitrogen for utilization by the plant during growth phase.

Auxins and Indol Acetic Acid (IAA) – FANTAC increases Indol Acetic Acid and other Auxins content resulting in better root growth, differentiation of phloem and xylem, cell growth, flowering and fruiting and promote femaleness in dioecious flowers.

Alpha Napthyl Acetic Acid (NAA) – FANTAC increases Alpha Napthyl Acetic Acid secretion which helps in better fruit setting and improves fruit quality like shape, size, weight, colour and helps in early ripening.

Cytokinins – FANTAC helps in production of cytokinins which stimulate cell division, enhance the period of stomata opening for increased photosynthetic activity, stimulate growth of lateral buds and stimulate chlorophyll synthesis.

Post harvest Pathology

Three major diseases of pomegranate caused by micro-organisms are grey mould rot, heart rot and penicillium rot. Grey-mould rot is caused by Botrytis cinerea. Decay usually starts at the calyx. As the disease progresses, the skin becomes light brown, tough and leathery. The pulp-bearing seeds disintegrate into a dark mass in advanced infections. Under moist conditions, a characteristic grey mycelium appears on the affected surface.

Heart rot is caused by *Aspergillus niger* and Alter aria sp. Affected fruits show a slightly abnormal skin colour but internally a mass of blackened arils becomes prominent. Usually there is a black line of decay extending from the calyx to the interior part of the fruit. The disease develops while the fruit is on the tree. Affected fruits can usually be detected by sorters and eliminated from the commercial pack. The rot due to Alternaria salani caused damage to fruits during storage and transit.

Penicillium rot, caused by P. expansum and other Penicillium spp., produces watery areas at the infection site followed by masses of blue or green spores. Infections in variably occur at skin breaks caused by cracking, mechanical injuries or insect punctures. Other fungus infect the same injured area and eventually overgrow the *Pencillium.* Other organisms causing decay in pomegranate fruit includes species of *Botrytis, Cladosporium, Phoma, Phomopsis, Rhizopus and Sphaceloma punicae.*

Dipping treatments with aqueous Topsin-M0. (1%) and Bavistin (0.05 to 0.1%) inhabited the growth of *Aspergillus niger.*

Infection of pomegranates with P. reticulosum causes the inner parts to develop grayish-green spore masses without any visible external symptoms. The possible toxicity of metabolic products of this fungus was investigated using *Paramecium caudatum* and white mice: the fungal metabolites caused the death of *P. caudatum* within 3-20 min. and of mice within 3-7 days after oral administration. *P. reticulosum* Birkinshaw has been known previously as a producer of mycotoxins. Further attention should be paid to its occurrence and possible hazards of its contamination of fruit, particularly that for juice production.

In view of the great importance of the pomegranate and considering its immense potential for cultivation under various adverse soil and agro-climatic conditions, more research efforts are needed for its inclusion in land-use management system, particularly on arid and semi-air lands. Apart from their well-established medicinal and curative properties, pomegranate fruits are also a rich source of minerals (calcium, magnesium and phosphorus). There are many wild forms of pomegranate grown in the sub-Himalayan areas which need proper exploitation for utilization into value-added products. Maturity indices based on physical criteria need to be established with respect to each variety. Pomegranates, being a non-climacteric fruit, have a tremendous potential for modified-atmosphere packaging (MAP) using various polymeric films which will not only retain fruit quality during storage but will also help in alleviating of chilling injury during refrigerated transport and storage. Therefore, an integrated approach on both production and post-harvest management using recent technologies on post-harvest handling, viz, individual shrink-wrapping, waxing, controlled-atmosphere (CA) storage coupled with judicious temperature management practices needs more attention for wide distribution of this delicious fruit in the global market.

(Courtesy S.K. Roy and D.P. Waskar MPKV, Rahuri)

PHYSIOLOGICAL DISORDERS

Internal breakdown of fruits

Unfortunately the production of good quality fruits of many varieties is set by development of serious malady known as internal breakdown or internal blackening, wherein, externally the fruit appears quite normal but its arils become soft having light creamy brown to dark, blackish brown colour. The incidence of internal breakdown after 150 days of anthesis and its intensity increases if the fruits are left on the tree up to 165 days. The incidence of browning increases with increase in weight of fruit. It is reported that TSS, acidity, ascorbic acid, total sugars, reducing sugars, calcium, phosphorus and the enzyme catalase were low whereas no-reducing sugars, starch, tannins, nitrogen, potassium, magnesium, boron, polyphone oxidize and peroxides enzymes were high in affected arils than the healthy ones.

Cracking of fruits

Cracking of fruits are believed to be due to a wide variation in soil moisture content. Accordingly high air temperature rise was found to be the cause of cracking. It amounted to 63 per cent in spring crop Jan-June, 34 per cent in winter crop (Oct-March) and only 9.5 per cent in rainy season crop (July-Dec) in variety Jodhapuri. It is observed that cracking of fruits as a varietals character and none of the cultivars was absolutely free of this trouble. Rind thickness and texture in various cultivars seem to be related to cracking. Cracked fruit shows reduction in weight of fruit and grains and volume of juice per fruit.

Ali Mohseni (2009) reported the introduction of aril decolouring of pomegranate and grain drying of pomegranate tree and difficulties created in pomegranate orchard in Iran. Iran is the main producer and exporter of pomegranate fruit in the world and during the recent years this fruit is exported to more than 35 countries in the world. According to latest formal statistics issued by Ministry of Jihad – e – Agriculture and IR Customs administration, total area of pomegranate orchard was 63,733 ha of which 56,329 ha with bearing trees and 7,404 ha with non bearing trees and total production exceeds 7,05,166 tonnes of which 24,389 tonnes exported earning US dollar 17.7 millions. As the area of new orchard and exporting of fruit is improving different issue of establishment, management, harvesting and post harvesting of orchards are more in attention recently. Pest and Disease in pomegranate orchard are the most important issue and Carob moth is a key pest of pomegranate for the last 40 years that causes serious losses since until not found any effective method to control it. This crop is playing important role in the revenue of many farmers in Iran and increasing the export of this fruit in recent years. There are two challenges of pomegranate orchards that causes losses are due to aril de-colouring of fruits and green drying of the trees.

Mahmoodi Tabar *et. al.*, (2009) study the aril paleness. A new physiological disorder that affect physical, chemical, anatomical and mineral nutrient properties of pomegranate fruit. Ferdows region in Iran where 2500 hectare pomegranate is grown with cv Shishe kabe Ferdows has been producing pomegranate high quality fruits by experts in 2001, aril paleness (Browning) was not observed by the gardeners. The present study describes the physical, chemical and anatomical and nutritional changes behind aril paleness disorder in pomegranate fruit Cv. Shishe kabe Ferdows. Fruit weight, seed percent, protein content and no significant different ($p = 0.05$) in intact and disordered fruits, juice percent, fruit and peel density and fruit firmness and decreased in disordered fruits (18.92%, 6.4%, 1.82%, 32.1% respectively) while peel percent, dry matter of juice, electrolyte of peel and aril increased in disordered fruits (16.46%, 8.94%, 13%, 18.56% respectively). Exposure of Aril paleness disorder caused the significant depletion in monomeric anthocyanin, phenolic compound, ascorbic acid, pectin content and titrable acidity (68.34%, 68.2%, 49.5%, 11.29% and 12.57% respectively). In disordered fruits polymeric anthocyanin acidity, total soluble solids and total sugars were significantly higher than intact fruits (2.8 – fold 11.61%, 10% and 17.33% respectively). Phosphorus, copper, manganese and sodium concentration showed no significant differences ($p = 0.05$) in intact and disordered fruits. The juice of disordered fruit showed significant reduction in concentration of Ca, Zn, and Fe, while K, Mg, B concentration increased significantly by exposure of disorder. Mineral nutrient showed no significant difference $p = 0.05$ in seed of intact and disorder fruits, with the exception of K increasing and Ca decreasing in disordered fruits. Anatomically vertical section of disordered arils showed that anthocyanin pigment degraded and fleshy tissue changed opaque. Hollow cavities and aerotic disordered cells were observed in fleshy tissue of disordered fruits.

Yazici and Kaynak (2006) studied the effect of Kaolin and shading treatment on sunburn on fruit of Hicaznar cultivar of pomegranate. Trees selected for shading were first covered with green covering material in the area, providing 35% shade on July 30th. Then effect of Kaolin application on sunburn were determined by applying 3% Kaolin with application at different periods that has been initiated at different dates with weekly intervals in the shaded and non-shaded areas [Period I (KI – July 30, August 13, August 20,) Period II (K II – August 7, August 13, August 20, August 27) Period III (K III – August 13, August 20, Augusut 27, and September 10)]. Application of Kaolin was the best method for preventing sunburn in fruits of Hicaznar pomegranate cultivar and increased soluble dry matter and red colour of fruit. The KI Kaolin application was the most effective for reducing sunburn in pomegranate fruits but it increased fruit cracking when humidity was high. Shading plus Kaolin was more effective for decreasing sunburn compared to shading treatment alone.

Mirdehghan and Rahemi (2006) studied the growth and development of pomegranate fruits in Iran. Seasonal changes of pomegranate fruit Malas Yazdi and its component, total soluble solids, pH, total phenolic compound, macro and micronutrients in the aril and peel were determined from 10 days after full bloom until harvest. The fruit show a sigmoid pattern of growth both on diameter or length basis, while fresh and dry weight of fruit accumulated linearly until harvest. Tss and pH in the arils increased while total acitiy decreased through out the growing season. Total phenolic level increased at early stage of growth both in peel and aril of fruits, but there after generally decreased during maturation and reached to 3.7 and 50.2 mg per gram of dry weight in aril and peel, respectively at harvest. The total amount of phenolics in peel was marked by higher than arils of pomegranate fruit. The concentration of most elements in aril and peel decreased during fruit growth and development. At harvest the relative order of concentration of macronutrients both in aril and peel was K > N > Ca > P > Mg > Na. The concentration of most micronutrients was greater in the aril than in the peel especially in the early season. The relative order of concentration of micronutrient in arils was B > Fe > Zn > Cu > Mn. So it is concluded that the nutritional contents in the edible parts of pomegranate fruit could be a good source of bio-active compounds and minerals.

Matuskovic and Micudova (2006) studied the practice with chemical mutagen (Natrium azide) on growth habit of pomegranate. The pomegranate seeds were in different concentration of *Natrium azide* (0, 3; 0,5, 0,7 m M). The control variant was composed with the seeds without influence of the chemical mutagen. From the motherly plants which was cultivated was propagated the collection of plants from every variant with the vegetal form. With cuttings which are further cultivated in the glass house. It is well known that every morphogenetic modifications induce a different habital deformities bizarre forms, aesthetical abnormal growth notabilities, they are always the object of interest. In the first phase the attention was towards leaf area, crown form, colour of flowers. Next was to incorporate the weeping or globular form which are required for modification with cuttings, the selection of plants individual is possibly for bonsai form of growing.

The next was on the vegetative method of propagation by green and woody cuttings with the aim to achieve the maximal propagation coefficient. The quantitative vegetal attributes of the juvenile phase, they inhibited qualitative attributes of generative phase. So the influence of mutagenic substance is extremely progressive for plant individuals formation, which are unusual on the habitual aspects.

Rezaian *et al.,* (2006) studied the recognition of Nutritional reasons for pomegranate colorlesss seed formation in the Ferdow region of Khorasan province in Iran, from this place fruits are exported to Japan. The most commercial variety of the regeion is Shishe cap which has shown white color phenomena at the first flowering stage. The lack of micronutrient deficiency such as Zinc and Iron. So three levels of spray were taken at 0, 0.3, 0.5% which will be sprayed twice a month's interval and before spray tree leaf samples were analysed.

Tripler and Harari (2006) studied the effect of maturation and ripening on pomegranate fruit quality. Fruit attractiveness is a function of both sensory and health promoting qualities. Pomegranate is grown in Arava valley in Israel (latitude 29°52'N; longitude 35°E') which is having desert climate suitable for early maturation, but fruits are of inferior quality in size, weight both external and internal colour.

Yazici and Kaynak (2006) studied the effect of air temperature, relative humidity and solar radiation on fruit surface temperature and sunburn damage in pomegranate cv. Hicaznar under arid and semi arid condition because of excessive heating sunburn is noticed. The present study was to know the effect of temperature, Relative humidity, solar radiation on fruit surface. The typical surface temperature of the fruit exhibited a typical symptoms of sunburn damage and no damage on ones were measured with thermocouples throughout the season. The sunburn is noticed with 41.0°C to 47.5°C depending on air temperature. Maximum Fruit surface Temperature (FST) exceeded 47.5°C in some days of July, August and September. Maximum daily temperature were highly correlated with maximum FST and it never exceeded 41 – 47.5°C in days when maximum air temperature was below 30°C. Mean solar radiation between 610 Wm^{-2} and 900 Wm^{-2} was highly correlated with maximum FST. The ratio of relative humidity around 70% and 80% showed inversely correlation with maximum FST.

Tumer *et al* (2006) studied the changes on the amount of phenolic compounds on maturation stages of some pomegranate varieties of Turkey. The eight phenolic compounds, gallic, chlorogenic, caffeic, catechin, protocatecuik, ferulic, P – coumaric and O cumaric acids were determined in juice seeds, peel of Hicaz, Silifke asisi and Cekirdeksiz VI during three maturation stage with HPLC. The amounts of phenolic compounds were changed according to varieties, maturation stage, fruit tissue and kinds of phenolic compounds. The highest phenolic compound were found in fruit peel, fruit juice, seed respectively. The amount of gallic acid, catechin and chlorogenic acids was respectively the highest in all varieties maturation stages and fruit tissue.

Yilmaz and Ozguven (2006) studied the effect of some plant nutrients. Gibberlic acid and pinolene treatment on the yield, fruit quality and cracking in pomegranate. Fruit cracking is a physiological disorder that occurs because of the different growth rates, between skin and flesh of fruits. To control fruit cracking various nutrients like boric acid, magnesium sulphate and potassium nitrate were applied in 1996-97 at two different times. The rates of fruit cracking were determined at harvest. Pomological analysis were conducted after harvest. The most effective treatment against fruit cracking was found to be gibberlic acid. It was observed that 150 and 200 mg L^{-1} GA_3 delayed the maturation of fruits and abscission of leaves.

Sheikh and Rao (2006) studied the effect of chemicals on fruit rind and aril colours in pomegranate cv. Ganesh. The pomegranate do not develop attractive rind colour, undesired rind colour affects the marketability of fruit. Spraying potassium dihydrogen orthophosphate (0.5%) or potassium nitrate (1%) effectively improved fruit colour to red and red yellow, the aril colour was either reddish or pink in colour by this treatment.

Sheikh and Rao (2006) studied the effect of chemicals on control of fruit cracking in pomegranate variety Ganesh. The fruit are prone to cracking, about 25 – 30% of the harvested fruits known to have cracks. Spraying of Boric acid reduced the cracking. The most effective treatment was 0.2%. Boric acid which also helped in marked increased in the harvest of healthy fruits.

Sheikh & Manjula (2009) studied the control of fruit cracking in pomegranate. Fruit cracking is a major problem in pomegranate which is mainly observed during drought period where the strengthened tissue develops in xylem, phloem, and loose their enlarging ability. After dry spell if sudden water supply is given, the meristamatic tissue because of differential growth ruptures. Hypertrophy of lenticels promoted greatly by retarded transpiration accompanied by plentiful water supply to the regions of hypertrophy. Heavy summer rains (April-May) causes fruit cracking of the plants were under severe water stress, because the skin get hardened and water is filled up in arils. During summer if 25% available soil moisture is maintained then fruit cracking will be less. Since calcium is cell binding material, spray of calcium chloride 1 kg/100 liter of water and calcium ammonium nitrate 2 kg/100 liter of water reduces fruit cracking. Dry heat accompanied by dry hot wind at the time of fruit ripening in pomegranate was also one of the reasons for fruit cracking. The rapid flesh growth temperature higher than 38°C combined with less than 60 per cent humidity favoured cracking. Sharp fluctuation in day and night temperature coupled with heavy irrigation after dry spell cause cracking. Nutrients like boron, zinc, calcium, copper, molybdenum, manganese potash involve in physiological process during fruit growth period and their deficiency cause cracking. Boron and copper helps in increasing the growth rate by stimulating enzymatic action in the peripheral tissue. Boron applications may probably help in translocation of sugar and synthesis of cell wall material and increases in methyl esterase activity. There is a direct relation with boron and pectic substance in the cell. Gibberlic acid @ 100 ppm in overcoming fruit cracking in pomegranate by increasing rind thickness.

Chapter 10

Plant Growth Substances

PLANT HORMONES: STATUS AND DEFINITIONS

Plant hormone has been defined as organic substance produced naturally in the higher plants, controlling growth or other physiological functions at a site remote from its place of production and active in minute amounts. To distinguish it from animal hormone. It was termed as phytohormone. However, has started a discussion about abolishing the name of plant growth substances as hormone. This idea was hotly debated at the XI International Conference of Plant Growth Substances. The crisis of the phytohormone theory is caused mainly by the following problem: a poor correlation was found between growth rate and endogenous hormone levels in developing plant. Further, unexpected results were frequently found with exogenously applied hormone, i.e., in some cases, promoters caused an inhibition that inhibited a promotion of growth. Plant growth regulators are defined as organic compounds other than nutrient which in small amount promotes/ inhibit or otherwise modify any physiological response in plants. The definitions adopted by the American Society of plant Physiology in 1954 are as follows:

Plant hormones are regulators produced by plants which in low concentration regulate a physiological plant processes.

Hormones usually move within plant from site of production to site of action. There are five classes of phytohormones: auxins, gibberellins cytokinins, abscistic acid and ethylene.

Auxins: These are organic substances which at low concentration (less than 0.001 M) promote growth along the longitudinal axis, when applied to shoot to plants freed as far as placing from their own inherent growth promoting substances.

Gibberellins: These substances are having gibbane ring skeleton capable of producing the same physiological responses as gibberellic acid that it must be active in specific gibberellin bio-assay. The gibberellins are phytohormones which are active in regulating dormancy, flowering, fruit setting and stimulating germination of seeds and extending growth of shoots.

Cytokinin: These are substances composed of hydrophilic group of higher specificity (adenine) and one lipophilic group of plant hormones having similar effects as those of GA in breaking the dormancy of a wide range of seeds and increases fruit set. These hormones mainly stimulate cell divisions and prevent chlorophyll degradation.

Various synthetic and natural plant growth regulators are listed below.

Names and structures of plant growth regulators.

Abbreviation or Common Name	Chemical Name or Structure	Major uses
ABA	Abscisic acid	Defoliant
ACC	1-aminocyclopropane-1-carboxylic acid	
Acifluorfen	5-2-Chloro-4-(trifluoromethly) phynexly-3-nitrobenzoic acid	GR Herb.
AVG	Amioethyoxy vinyl glycine.	
Alachlor	2-chloro-2', 6'-diethyl-N-(methoxymethyl) acetanilide	Herb.
Ancymidol	a-cyclopropyl-4-methoxy-a-(pyrimidin-5-yl) benzyl alcohol	GR
Atazine	2-chloro-4 (ethylamino)-6-(isopropylamino)-1,3,5-triazine	
BSAA	(Benzo-b-selenienyl)-3-acetic acid	DB
Barban Brassinolide	4-chloro-2-ynyl-3-chlorocarbanilalte	Herb.
Chlormequat, CCC	2-Chloroethyl-trimethyl ammonium chloride	LP,SR
Cycloheximide	3-{2-(3,5-dimethyl-2-oxocyclohexyl)-2 hydroxyethyl}-glutarimide	AS
3-CPA	2-(3-chlorophenoxy) propionic acid	FT
4-CPA	4-chlorophenoxy-acetic acid	FT
CPPU	N-(2-chloro-4-pyridyl)-N'-phenylurea	Herb.
2,4-D	(2,4-dichlorophenoxy) acetic acid	Herbs
Daminozide SADH	Succinic acid-2,2dimethyl hydrazine	GR
Dichlopentezol	(E) - 1(2,4-dichlorophenyl)-4-4-dimethyl-2(1,24-triazol-1-yl)-1-penten-3-ol	GR Fung
Dikegulac	Sodium-2,3-4,6-di,I-isopropylidene-?-L-xylo-2-hexulofurasonate	GR
Diquat	6,7-dihydrodiprido pyrazinediium dibromide	FS
Diuron (DCMU)	3-(3,4-dichlorophyeyl)1,1-dimethylurea	
DPX-1840	2-(4-methoxphenyl)-3, 3a-hydro-8H-pyraxolo {5,1a} isoindol-8-one	GR
Etacelasil	3-chloroethyl-tris-(2-methoxyethoxy)silan	FL

Ethephon	2-chloroethylphosphonic acid	SR
Glyphosate	N-N-bis(phosphonomethyl) glycine	SR Herb
Glyphosine	N,N-bix (phosphonomethyl)glycine	SR
GA_3	Gibberellic acid	GE
Kinetin	6-Furfury laminopurine	-DB
IAA	Indole-3-acetic acid	Enlarge
IBA	Indole-3-butyric acid	Enlarge
Maleic hydrazide	1,2-dihydro-3, 6-pyridazinedione	Herb, GR
Mefluidide	N-2,4-dimethyl-5- (trifluoro methyl) sulfonylamino-phenyl acetamide	GR Herb
Mepiquat chloride	N,B-dimethyl piperidiunium chloride	GR
Metolachlor	2-chloro-N-(2-ethyl-6-methylphenyl)N-(2-methoxyl-1-methylethyl) acetamide	Herb
Metribuzin	4-amino-6-tert-butyl-4,5-dihydro-3-(methylthio)-12,4-triazine-5-one	Herbs
Monuron	3-(p-chloro-phenyl)-1,1-dimethylurea	Herbs
NAA	1-naphthalene acetic acid	FT GR
NOXA BNOA	2-nathpthhalenyloxyacetic acid	GS
Paclobutrazol	(2RS, 3RS) -1-(4-chlorophenyl)-4, 4-dimethyl-2-(1,2,4-triazol-1-yl)-pentan-3-olk	Gr
Paraquat	1,1-dimethyl-4,4-bipyridinium dichloride	Herb
Pretilachlor	2-chloro-2', 6'-diethyl-N-(2-propoxyethyl)acetanilide	Herb.
Propachlor	2-chloro-N-isopropylacetanilide	
Propazine	2-chloro-4-,-6,bix(isopropyl-amino)-1,3,5-triazine	
Release	5-chloro-3-methyl-4-nitro-1H-pyrazole	As
Terbutryn	2-(tert-butylamino)-4-(ethylamino-6(methylthio)-5 triazine p-hydroxyphenyl)-acetic acid	
2,4,5-T	(2,4,5-trichlorophenoxy) acetic acid	Herb
Tetcyclacis	5,(-4chlorophenyl) -3,4,5,9,10-pentaaza-terna-cyclo(5,410^{2,6}, 0^{8,11}) dodeca-3,9diene	
Thidiazuron	N-phenyl-N'-1,2,3-thiadiazol-5-ylurea	CD GR
TIBA	2,3,5-triiodo-benezoic acid	GR
TRIA	1-hydroxy-triacontane triacontanol	GS
Triadimefon	1-(chlorophenoxy)-3,3-diamethyl-1-(1,2,4-triazol-1yl)-2 -butanone	GR

Triadimenol	3-(4-chlorophenoxy)-(1,1-dimethylethyl)-1H-1,2,4-triazole1-ethanol	GR Fiung
Triapenthenol	(E) (RS) 1-cyclohexyl 4,4-dimethyl 2 -(1 H 1,2, 4-triazol-1-yl) pent -1-en-3-ol	
Uniconazol	(E)0(4-chlorophenyl)-4-dimethyl-2-(1,2,4-triazol-1-ly)-1-penten -3-ol	GR Fung.

Abbreviations: AD= Apical-dominance reducer; AS=Absission stimulant: CD =Cotton defoliant; DB = Dormany breaker; Enlarg=Plant cell enlarger; FL = Fruit loosener; FS = Flowering suppressant, in sugarcane; FT=Fruit thinner; Fung = Fngicide; Gamet = Gametocide; GF = Grape enlarge; GR = Growth retardant; GS = Growth stimulant; Herb = Herbicide; LP = Lodging prevent or; PF = Pineapple floweing agent; SR = Sugarcane ripener.

Abscisic Acid (ABA): ABA is a naturally occurring sesquiterpene which regulate plant growth and metabolism in various ways and have been detected in nearly all plants. It is involved in the abscission of plant organs, induction and vegetative buds, in regulation of fruits' repeining and generally in reduction of growth.

Ethylene: It is the only gaseous hydrocarbon hormone which plays an important role in the ripening of fruits, inhibition of root growth, abscission and other growth processes. Unlike the other harmones, ABA and ethylene are not discovered through any interaction with fungi.

List of hormones and their physiological responses

Effects	Auxin	GA	CK	ABA	Ethylene
Cell elongation	+	+	+	-	-
Cell division	+		+		-
Apical dominance	+		-		
Leaf growth					
In breadth		+			
Chlorophyll formation		+			
Vein elongation	+				
Leaflet movement	+				
Root formation	+				
Fiber growth in cotton seed	+	+	-		
Abscission	-	-		+	+
Sex expression					
Maleness	-	+		-	
Femaleness	+		+		
Fruit growth and division	+	+	+	+	+
Xylem formation	+	+			
Flower	+	+	+	-	-
Seed dormany		+	+	+	-
Bud dormany		+	+	+	-
Senescence	-	-	+	-	
Tropism	+	+		+	-

+Promote, -inhibit,

HOW HORMONE ACTS?

When a hormone acts upon a responsive plant system, it occurs, enters into some direct and molecular interactions, which results eventually in manifestation of measurable effects (bio-chemical or physiological responses). However, there are two aspects of hormone action.

1. Mechanism of action: **The direct and specific molecular interaction, and**
2. Mode of action: **The succeeding series of steps (events and processes) which results in the measurable bio-chemical or physiological responses.**

The two-term mechanisms and Mode of action should be taken as different terms rather than synonyms. The mechanism by which minute amounts of plant hormones, relatively simple organic compounds, can do dramatically controlled growth is one of the pressing and challenging problems in plant physiology. The progress in this field has been quite spectacular but still it needs a proper synthesis and analysis of various concepts given in this regards.

In the previous years numerous physiological and bio-chemical responses to plant hormones have been observed which have a bearing on the mechanism of hormone action. Some of the responses have been enumerated they are as follows:

1. Increased plasticity of the shoot and elasticity of the root cell wall.
2. Increased permeability of water.
3. The enhanced capacity to retain water taken up.
4. The active uptake of water and solute apparently independent of osmotic forces and occurring even against an osmotic gradient.
5. A decrease in protoplasmic viscosity.
6. An accelerated rate of respiration and cyclosis.
7. More rapid synthesis of protein, with lowered levels of free amino acids.
8. The increase of monosaccharide at the expense of reserve polysaccharide.
9. An increment of (intermicellar) wall pectin and of cellulose and a stimulation of alcohol and malic dehydrogenises, catalane, phosphates and ascorbic acid oxides.
10. These large varieties of effects which hormone evokes depict the multiplicity of sites.

An understanding of hormone action is dependent on location of sites of action. We are beginning to learn that the site of action of hormones is close to gene. Usually two sites of plant hormone action at the sub-cellular/molecular level are discussed; an early (primary) effect at the cell membrane and a subsequent (secondary) effect at the genetic level.

The value added products like pomegranate juice, jelly, anardhna, anarrub, rind powder etc., are prepared from pomegranate fruits. The 1000 g of pomegranate wine can be prepared through fermentation technology by using suitable wine yeast, (Ezyhrinciew, 1996; Probaglav et al, 1983; Shivanna 1997; singhnagi and Manirekar 1975; Suresh and Ethiraj 1968 Suresh et al. 1985). Wine is a fermented beverage from fresh fruits composed of water, alcohol, pigments esters, vitamins, carbohydrates, minerals, acids, tannins with medicinal and theropeutic value.

Anti-transpirant

It is a material which is applied to transplanting plant surface with an aim to reducing the water loss from the plant (mainly leaf)

Approximately 99 per cent of water taken by the plant through roots is transpired to the atmosphere through the stomatal spores of leaves. The stomata bearing leaf surfaces are most important sites for the anti-transpirants applications. The types of anti-transpirants are described below:

Film forming types

Plastic film forming anti-transpirants are more permeable to CO_2 than H_2O under water stress e.g. higher alcohol, ethyl achohol, hexodecasol S-789 Tag-9)

Metabolic inhibitors

Decreases more transpiration (50%) but decreases the photosynthesis to a lesser extent by 35 per cent by closing stomata under moisture stress. E.g. phenyl methyl acetate (PMA) (fungicide) are most widely used as anti-transpirants.

Reflectant types

Reflecting coating raises the light saturation level and increases light compensation points of all species. Improves water use efficiency with increased radiation e.g. 2 per cent Kaolnite spray.

Plant growth regulators

These are plant growth regulators which close the stomata under moisture stress e.g. ABA (abscissic acid) and 2,4,5-T.

BAHAR TREATMENT

Bahar treatment means giving rest to the plants holding of water. In other words it it artificial stress creation within the plant by no supply of water and nutrients.

C:N ratio is normally 28-2 parts of C to 1 part of N. So the ratio is 28:1.

It is known and analysed that during flowering stage the ratio comes to 14:1 which means that there is depletion of CHO and not accumulation. The lower CHO means relatively high N and it is this relatively higher N that gives rise to flowering.

The stress period depends upon the type of soil or heavy soil, more the stress on light soil, lesser the stress period. Thus nutrient concentration is maintained

BAHAR TREATMENT INCLUDES

1. Water withholding
2. Nutrition and fertilizer use.
3. Change of soil in the root zone.

As the stress during bahar treatment are being resorted to there is no need of addition of nutrients. The initial nutrition plays role in giving profuse flowering.

If the tree has more concentration of balanced nutrients more is the flowering.

In ring and basin systems of irrigation, every year we add lots of nutrients out of which the portion of soil replaced by ring soil. The reasons behind that is fertilizers are added to the root zone, which are used by the plants and soils around that portion which get depleted. Thus while giving bahar treatment the soils of these portions are interchanged. Hence every year we add enriched soil to the feeding zone and depleted soil is put into the ring zone. This is how soil fertility can be maintained.

Pruning of unwanted roots is also done through bahar treatment.

Hormones are organic chemical messengers that are produced in one tissue and located to another where they produce their specific effect, and regulate plant growth and development.

The term hormone comes from Greek hormone and means "to excite" this might give impression that all hormones of plants are stimulatory to growth. Many hormones inhibit growth. Many synthetic compounds mimic the effect of hormones. Hence the term plant growth substance or growth regulator is used for any hormone-like compound, natural or synthetic. The term hormone is restricted to natural products only.

Characteristics of Plant Hormones

i) Plant hormones are produced most abundantly in the actively growing parts of the plant body such as apical meristems of the shoot and root, young growing leaves, or developing seeds of fruits.

ii) They are trasnsported both through the vascular tissues and through non-vascular tissues such as parenchyma. The transport is in any one particular direction, hence it is polar transport. The direction is generally basipetal, i.e., from the tip to the base of an organ. It is probably acropetal, i.e., from the base to the tip in some.

iii) Only minute amounts of the substance are needed to influence a physiological process. They act in less than one-thousandth molar and often in one-millionth molar range.

iv) Unlike animal hormones, it is difficult to separate the site of production from the site of action.

v) Plant hormones may be stimulatory or inhibitory depending on the concentrations, i.e., they can have a positive or promoting effect or can have a negative or inhibiting effect.

vi) The same hormone may produce different effects, i.e., promote some processes, inhibit some others and may not produce any effect on others.

vii) Different hormones interact in complex ways to control a particular physiological process (growth, development or metabolism).

Kinds of Plant Hormones

Just as a machine needs both an accelerator and a brake for effective and precise control, and also the growth systems of plants need growth-promoting substances and growth-inhibiting substances. There are five main groups of plant growth substances—three of them (auxins, gibberellins and cytokinins) are growth promoters and two (abscisic acid and ethylene) are growth inhibitors.

Growth Promoters

The Auxins

The auxins were the first plant growth substances to be identified. Naturally occurring auxins are weak organic acids. The acidic group is bound to the end of a side chain attached to a ring compound. The most common naturally occurring auxin is identified as indole acetic acid (IAA).

Natural and Synthetic Auxins

i) Naturally Occurring Auxins: The most common naturally occurring auxin has been identified as indoleacetic acid (abbreviated as UAA). It is derived from the amino acid tryptophan. It is produced mainly in the shoot apical meristems, young leaves and seeds. A very small amount is also produced in the root tips.

ii) Synthetic Auxins: Many substances similar to auxins have been synthesized in the laboratory. These include two phenoxyacetic acids: 2,4-D (2,4-dichlorophenoxyacetic acid) and 2,4,5-T (2,4,5 trichlorophenoxyacetic acid); IBA (indole butyric acid); NAA (naphthalene acetic acid) and two benzoic acids: 2,3,6-trichlorobenzoic acid (TCBA) and 2,3,5-triiodobenzoic acid (TIBA) are the others.

Effects of Auxins

Auxins influence a number of activities of plants. The following are some of them:

i) **Cell Division and Differentiation:** The vascular combine is the site of response to auxins. Cell division in the cambium is stimulated, so also the differentiation of the newly produced cells into xylem and phloem elements. If apical mersitem is destroyed the basipetal flow of auxin stops and vascular tissue differentiation is inhibited.

ii) **Cell elongation:** Auxins stimulate elongation of cells in the young internodes just below the apical meristem. Auxins possibly act by actively transporting protons across the cell membrane and thereby activating the enzymes that weaken the bonds holding the cellulose molecules of the cell wall.

iii) **Apical Dominance:** The growing apical meristem of a plant prevents the growth of lateral buds. This is called apical dominance. If the apical meristem is cut off, axillary buds soon grow to produce lateral branches. If an auxin is applied to the cut surface of the stem, the buds will continue to be dormant. If a potato tuber is treated with auxin, the lateral buds ("eyes") will remain dormant. Hence an auxin maintains dormancy. Once the buds begin to grow, auxins stimulate their growth.

iv) **Root Initiation:** Auxins stimulate the growth of adventitious roots. If cut ends of stems are dipped in dilute solution of auxin, roots develop at the cut ends. New plants may be propagated this way.

v) **Flower Initiation:** Application of low concentrations of 2,4,-D and NAA initiate flowering in pine apple.

vi) **Control of Abscission:** Abscission is a process by which leaves and fruits drop off from a tree by the formation of a weak layer of cells called the abscission layer in their stalk. Young leaves and fruits are firmly attached to the stem as long as they produce auxins. When the quantity of auxin produced becomes less, the abscission layer develops at the base of the petiole or the fruit stalk. Soon the petiole or the fruit stalk breaks at this point and it falls to the ground.

vii) **Prevention of Premature Withering of Flowers and Fruits:** The spraying of 2,4-D,IAA and IBA prevents premature (pre-harvest) withering of flowers and fruits of orchard plants (citrus, pear and apple).

viii) **Induction of Parthenocarypy:** A fruit develops from the ovary of a flower. Normally a fruit develops only after pollination and fertilization. When auxins are applied to immature flowers, the ovary is stimulated to develop into fruit even before pollination. Such fruits develop without pollination or fertilization and are called parthenocarpic fruits. The developing embryo or seed itself is the source of auxin for the development of the fruit. If the seeds are removed during development, the fruit stops growing.

ix) **Sex Expression:** Spraying of auxins causes cucurbit plants increase the number of female flowers.

x) **Effect on Tropisms:** A growth response to light is phototropism. If a growing plant is illuminated by light on one side, it bends towards light; the reason for this is the inactivation of auxins on the illuminated side by light (photoinactivation). This produces unequal concentrations of auxins, higher on the shaded side, and lower on the illuminated side. Differential growth occurs and the shoot bends.

   Unequal concentrations of auxins and the inhibitory action of auxins on roots is given as the cause for positive geo-tropism (now called gravitropism) in roots.

xi) **As Herbicides:** the synthetic auxins 2,4-D and 2,4,5-T are selectively toxic to unwanted plants called weeds, growing among crop plants, cereals and lawn grass.

Hence they act as weed-killers.

The Gibberellins:

Effects of Gibberellins

i) **Elongation:** When gibberellins are applied to mutant dwarf plants which have lost their capacity to synthesize gibberellins, they grow quite tall and become indistinguishable from the normal tall plants. Application of gibberellins to normal plants may also produce elongation. Too much gibberllin causes the stem to become long and thin with very few branches and pale stems.

ii) **Bolting and Flowering:** Biennial plants such as cauliflower and cabbage form leaf-rosette before they produce flowers. Flowers are produced under certain specific photoperiodic conditions. Application of gibberellins causes stem elongation (upto 4-5 meters) and flowering in these plants. This phenomenon is called bolting. In a normal plant bolting occurs only in the second year of growth. Gibberellins can induce bolting in the first year itself.

iii) **Sex Expression:** It has been observed that gibberellins influence the formation of male or staminate flowers in cucumbers and squashes.

iv) **Breaking Dormancy:** Gibbrllins break the dormancy periods of seeds, stimulating them to germinate. In the seeds of the Graminease, there is specialized layer of cells called the aleurone layer just below the seed coat. This is rich in protein. Just before germination, when the seeds absorb water, gibberllin is produced which diffuses into the aleurone layer. In response to gibberllin the aleurone layer produces several enzymes which speed up digestion of reserve food, weakens the seed coat, thus making it easy for the root to emerge out.

v) **Parthenocarpy:** Gibberllins can induce the formation of parthenocarpic fruits and are more potent than auxins in this context.

vi) **Hybrid Vigour:** Gibberllins play a role in hybrid vigour. An analysis of gibberellins of hybrid and inbred strains of plants, in 1988 showed that hybrids produce higher levels of gibberllins. When additional gibberellins are applied to inbred plants, their growth also approaches that of hybrid plants.

Cytokinins produce the following effects:

i) **Cell Divison:** It has been established in various tissue culture experiments, the most important function is stimulating cell division. It also promotes cell elongation.

ii) **Organ Formation:** Cytokinins also plays a role in organ formation, i.e., morphogenesis of roots and shoots. Varying concentrations of cytokinin/auxin ratios can induce development of roots/shoots or both.

iii) **Lateral Bud Development:** When cytokinin is applied to a non-growing lateral bud, it is stimulated to grow.

iv) **Chloroplast Development and Chlorophyll Synthesis:** Application of cytokinins to etiolated leaves, enhances development of chloroplast, especially increases the rate of chlorophyll synthesis.

v) **Delay in Senescence:** Along with auxins, cytokinins prevent premature aging or senescence by stabilizing proteins and chlorophylls.

vi) **Breaking Seed Dormancy:** Cytokinins are known to break seed dormancy and induce germination.

Growth Inhibitors

Abscisic Acid:

The best known growth inhibitor is abscisic acid (ABA). The name is derived from the ability of the substance to promote abscission, a discovery made by F.T. Addicott. It is synthesized in fruits and leaves and has important effects on growth and development.

i) **Abscission:** It promotes senescence of leaves. It play a role in formation of abscission layer at the base of a leaf flower on fruit. The abscission layer breaks to cause the fall of leaves and fruits.

ii) **Dormancy:** It suppresses elongation of shoots, inducing the formation of dormant buds.

iii) **Stomatal Closure:** It functions as a stress hormone and helps to cope with drought by stimulating closure of stomata

iv) **Defence against Salt and Cold Stress:** When plants are under stress in conditions of inadequate water supply, salinity and chilling and freezing temperatures. ABA level increases in plants, suggesting that it helps plants to overcome the 'stress' to some extent.

v) **Embryo Development in Seeds:** ABA inhibits development of embryos in seeds.

Ethylene

Ethylene is the only gaseous or volatile plant hormone. It is a product of metabolism of the amino acid methionine and is produced in nodes in stems, ripening fruits and senescence tissues. It is released from most plant organs of a varying concentration, especially from ripening fruits.

i) **Fruit Ripening:** Ethylene is a fruit ripening agent. It is released in large quantities during the ripening phase called climacteric. When complex carbohydrates are broken down into simple sugars, chlorophyll breaks down, cell walls become soft and volatile aromatic substances are released.

ii) **Triple Response:** It inhibits elongation of stems, increases stem thickening and promotes horizontal growth habit.

iii) **Senescence:** It promotes senescence of flowers and leaves and their abscission.

iv) **Additional Effects:** It initiates formation of root hairs, latex production and the formation of parenchyma in submerged root and stems.

Plant Hormones and some Known Effects

SI. No.	Hormone	Effects
1	Auxin	Promotes cell elongation; formation of roots in cutting; apical dominance; fruits and maturation; xylem differentiation; tropism
2	Gibberllin	Promotes stem elongation (in dwarfs); involved in flowering; release of dormancy of buds and seeds; stimulates growth of pollen tube during fertilization.
3	Cytokinin	Cell division; bud embryo development; slows down leaf aging.
4	Abscisic acid	Promotes stomatal closure; initiates bud dormancy; probably not involved in abscission
5.	Ethylene	Promotes fruit ripening; promotes abscission of leaves, flowers and fruits; initiates root hair, latex precaution and formation of parenchyma in submerged roots and stems.

Application of Phytohormores in Agriculture and Horiticulture

Plant growth substances especially synthetic auxins, and to some extent, ethylene have found immense uses in agriculture and horticulture. Some are mentioned below:

i) **Rooting of Cutting:** When cuttings of woody plants are to be planted, it is essential that they sprout adventitious roots when planted. If the cut end is dipped in a solution of a synthetic auxin like NAA (naphthalene acetic acid,) roots arise quite quickly. A successful development of new plants can thus be achieved.

ii) **Fruits-setting** NAA and IBA have been extensively used to improve natural fruit-setting in trees in orchards.

iii) **Flower Initition:** NAA, 2,4-D (2,4-dichlorophenoxy acetic acid), when sprayed on leaves (foliar sprays) in pineapple, apple, and pears helps to cause flowering and also fruit-ripening.

iv) **Dormancy in Potato Tubers:** To prevent the lateral buds ("eyes") of potatoes from sprouting during storage, agriculturists use sprays of IBA and methyl esters of NAA. With this treatment potatoes can be stored for as long as three years.

v) **Premature Fruit Fall;** 2,4-D can counter the effects of abscisic acid (that promotes the dropping of fruit from trees). Application of a spray to pear and apple trees is a practice to make the trees retain and ripen more of their fruits.

vi) **Weed-Killing:** 2,4-D; 2, 4, 5-T; 2, 3, 6-trichlorobenozoic acid (TCBA) and 2, 3 5, -triiodobenzoic acid (TIBA) are used as selective weed-killers. Most weeds are fast growing dicotyledons; they are undesrirable. In a field of monocot crops, if the weed-killers are sprayed in certain dosage, the weeds grow abnormally, grossly deformed, and finally die. Monocots are not visibly, affected. 2,4,5-T was particularly effective against woody perennials of rough pastures. It has been banned since it contains traces of toxic dioxin, which can cause foetal abnormalities, a severe form of acne (chlorachne) and cancer.

A mixture of 2,4-D and 2,4,5-T was sprayed from aeroplane is Vietnam War by USA to defoliate (-to remove leaves) vast areas of countryside. The contamination was detected in drinking water and fish. It was also the reason for abnormal development of some babies. Later, this programme, called 'Agent Orange', was given up.

vii) **Fruit Ripening:** Ethylene gas is released slowly from a chemical called ethopan, (commercially available as Ethrel). It is 2-chlorethyl phosphoric acid which rapidly breaks down in water at alkaline pH to release ethane. It stimulates the production of enzymes that promote fruit-ripening. If fruits in tropics are to to be sent to far off places, fruit growers can pick citrus fruits, bananas and pineapple when they are still green. They can be ripened by application of ethopan after they reach their destination.

Practical utility:

1) **Defoliation:** Ethrel spray @ 2.5ml/liter of water helps in defoliation in pomegranate.
2) **Flower thinning:** Spraying @ NAA 250 ppm helps in thinning of flowers.
3) **Fruit size:** Spraying GA @ 20ppm at lime size in pomegranate helps in improving fruit size.

Sheik and Rao (1999) the pomegranate plant bear flowers in abundance, However not all flowers are productive flowers, the flower may comprise staminate, intermediate and Hermophrodite flowers. Excess fruit set affect fruit size and yield. It is necessary to thin flowers/fruits to improve the fruit size. Since hand thinning is laborious and cumbersome, chemical thinning of flowers, fruits by NAA, Ethrel spray of different concentration cause thinning in the ratio of 1:2. 1:3 and 1:3 to 1:5.

Chapter 11

Post Harvest Technology

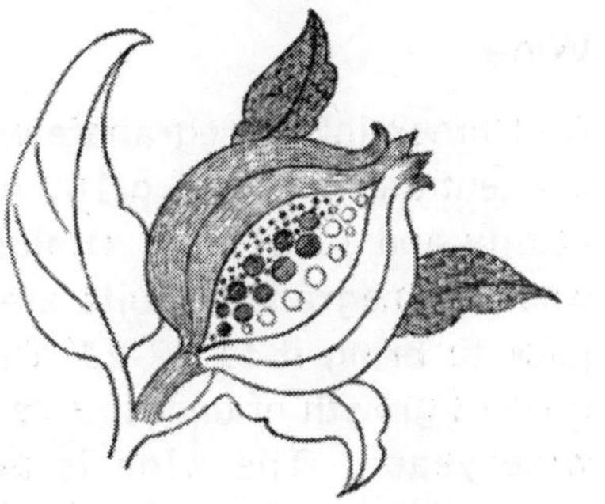

Sound pomegranate fruits fetch a fairly good price in the market and are hardly used for processing purpose. Hence, limited information is available on the utilization of pomegranate fruits for processed products (Elyatem and Kader, 1984). During seasonal glut, the prices of fruits are fairly low. Cracking of pomegranate fruits hampers their marketability. Such fruits can be utilized for processing. Pomegranate fruits with large juice grains can give an attractive coloured (purplish red) juice with mild acid sweet taste and low tannin content (less than 0.25 %) are the desirable characteristics for processing of pomegranate.

Juice

Pomegranate juice makes a delicious drink (Gabunia et al., 1984). The juice may be extracted from whole fruits or from grains. On whole fruit basis, the yield of juice is about 42 per cent, while from grains the yield is about 70 per cent (Siddappa, 1943). For extraction of juice, the fruits are cut into quarters and pressed under moderate pressure in a rack and cloth hydraulic press to recover 36.7-41 per cent juice (Saxena *et al.,* 1987). The juice can also be extracted using spiral type screw press without crushing the seeds. Pomegranate juice is also extracted from the seeds by pressing in a basket press. The juice obtained by pressing grains contains less tannin (0.12 %) than that obtained from whole fruits (0.175%) Siddappa, (1943). The excess tannins in the juice are precipitated by gelatin. The juice is clarified by pectinase enzyme or heating in a flash pasteurizer at 79-82° C, cooling, settling for 24h, racking up and filtering. The clear juice can be preserved by treatment or by using chemical preservatives. The juice may be preserved by addition of 600 ppm sodium benzoate (Siddappa, 1943). The use of SO_2 or potassium metabisulphite is not recommended for pomegranate because of loss of its natural colour due to the bleaching action of SO_2. The chemical composition of pomegranate juice has been reported by several workers. The juice has been reported to contain 16 to 17 per cent total solids (Veres, 1976) and 0.81 to 1.23 per cent acidity as citric acid (Chavan *et al.,* 1985). A technology for obtaining canned pomegranate beverage is described (Gabunia, 1984).

Wine

A commercial pomegranate wine made in California was reported to contain 11.9 per cent alcohol, 8.85 g/100 g total acidity (as citric acid), 0.07 g/100 ml volatile acidity and 11.5^0 Brix (Kainsa and Gupta, 1979). For preparation of wine, the whole pomegranate fruits are pressed without crushing. Sugar is added to the juice to bring it to $22\text{-}23^0$ Brix. Potassium meta bisulphite is added to juice to prevent growth of undesirable microorganisms. The juice is fermented with starter wine yeast. The wine is aged and finished in the same manner as in red pomegranate wine. Research carried out in our laboratories have shown that good quality wine can be prepared from pomegranate juice (Adsule *et al.,* 1992). The rate of fermentation of pomegranate juice was, however, slower than that of grape juice.

Jelly and Anar-rub

An attractive jelly can be prepared from pomegranate juice (Saxena *et al.,* 1979). A product known as anar rub with fairly good keeping quality can be made by concentrating pomegranate juice by adding sugar and heating the mixture on slow fire for long period. The finished product has thick consistency and contains 70-75 per cent total solids (Siddappa, 1943).

Anardana

Pomegranate seeds, particularly of sour types, can be dried and sold as anardana. It is used as an acidulant in place of tamarind or dried green mango for cooking purposes in North India, in Indian style curries, chutney and other culinary preparations (Saxena *et al.*, 1987). The anardana has high acidity (Pruthi and Saxena, 1984) ranging from 7.8-15.4 per cent. Pruthi and Saxena (1984) recommended the use of cross flow driers to get a uniform, hygienic, good quality anardana. The use of solar drier for dehydration of anardana has been reported (Saxena et al., 1987).

Rind Powder

This can be used as tooth powder, and in medicine and cosmetic industries. A dry powder has been successfully prepared from pomegranate skin. Various steps in the preparation of pomegranate skin powder have been out lined in fig. 3. The recovery of powder was found to be 13.3. per cent (* haven, *et al.,* 1985). This is an excellent source of B-carotene, potassium, phosphorus and calcium (Table). The powder contained 16.5 per cent polyphones and 5.35 per cent mineral matter.

Table: Chemical and mineral composition of pomegranate fruit and rind powder

Constituent	Fruit	Rind Powder
Moisture (%)	78.00	8.40
Protein (%)	1.60	0.94
Total sugars (%)	14.60	3.20
Ascorbic acid (mg/100g)	16.00	2.40
B-carotne (ppm)	-	12.52
Ash (%)	0.70	5.35
Polyphones (%)	-	16.50
Acidity (%)	0.58	4.13
Calcium (mg/100g)	10.00	35.90
Phosphorus (mg/100/g)	70.00	1100.00
Magnesium (mg/100g)	44.00	140.00
Potassium (mg/100/g)	133.00	1550.00
Sodium (mg/100g)	0.90	40.00
Iron (mg/100g)	1.89	30.00
Zinc (mg/100g)	0.82	30.00
Manganese (mg/100g)	0.77	2.00
Copper (mg/100g)	0.34	1.00

Selection of mature healthy pomegranate fruits

Washing

Separation of seeds and rind

Rind

FIG. 3 PREPARATION OF POMEGRANATE RIND POWDER

Bignami *et al.*, (2006) reported regarding collection, conservation and valorization of Traditional Fruit Trees varieties for NICHE Market Development. Therefore, investigation, collection and characterization of local genotypes in Lazio region on Pomegranate was performed with a special focus on areas with major agronomical, historical and cultural interest. In the surroundings of Monte Sen Biagio some old lots with cultivated pomegranate plants were found sometimes in poor conservation condition or semi-abandoned. The plants are sometimes grown in specialized lots or more frequently, in rows along canals that divided the lots. Its position nearby the ditch was useful to strengthen it, as well as to provide the plant with water. In different areas from Formia, Gaeta and Monte Sen Biagio countries six ecotypes of Pomegranate were collected; differences of carpological characters such as aril colour and fruit size, were observed. Descriptors lists were filled in, and some basic qualitative analysis were performed, which include soluble solid contents, titrable acidity and juice pH, plants collected were propagated by cuttings and then planted in a collection at the Experimental Farm of Universita della Tuscia, in viterbo. The collected varieties belong to pomegranate typology with medium acidity and good sugar content.

Rymon, (2006) examined the variation in the composition of pomegranate production costs in four countries viz., Australia, Israel, Spain and USA, and explained the similarities and dissimilarities among the countries. The comparison runs along two main tracks; the investment period and a typical full bearing year, say year full. To overcome the technical difficulties of making a valid inter-country comparison (e.g., different currencies) cost composition is expressed in proportions (percents), rather than in nominal figures in each country. Since the studies analyze actual cases, they do not fit in what an experiment list would regard as a comparable multi-country situation in which most of the parameters are maintained at a fixed level while the variable of interest are allowed to change. For example, one study focuses on production on poor soil, with all its technical implications. The comparison highlights a striking variation of the labour input; where labour is relatively expensive it represents the major expense and creates a strong incentive to search for an alternative i.e., development of machinery and mechanical devices to substitute for costly labour. A major limitation of this analysis is its lack of comparable yield data. Thus we cannot relate the cost of outputs. A specifically designed study carried out in several countries perhaps supported by an International entity (e.g., FAO, UNDP, IBRD or others), can produce the data for a valid comprehensive analysis.

Khoshrow *et al.,* (2006), conducted survey in Iran revealed that, in pomegranate industry, quality evaluation is still carried out manually by trained Inspectors, which is tedious and laborious. In order to develop an automated system for pomegranate quality evaluation, image processing techniques can be combined with mechanical and instrumental devices to replace human manipulative effort in the performance of the process. In such a system, Image processing is of central role which controls the operations of the machinery. The aim of this research is sorting pomegranates based on such external features as colour and surface defects. RGB (Red Green Blu) and HIS (Hiu saturation diseased) pomegranates. To analyze pomegranates for shape classification, some features i.e., area ratio, aspect ratio, circulantly, compactness and roundness extracted from images, are in the focus of this research.

Hadj Salem *et al.,* (2006) says that Tunisian pomegranate germplasm is very diverse containing many cultivars with sweet fruits and others with acid ones. Production of juice and other physio-chemical characteristics were studied for six acid cultivars and compared with two sweet ones. Juice proportion varied from 50% to 75% of the fruits edible part. Cultivar Garoussi 2 produced the less coloured and distributed juice, white Zaghouani 1 gave the cleanest juice with highest antocyanin content. Total soluble solids content varied from 141 g/L, for Garsi 1, to 170, 8g/L for Gabsi 1. Acid fruits produced abundant juice with, relatively high content of potassium, sodium and calcium. Thus, highest titrable acidity (20,6 g/L) and total salts content (3,15 g/L) were registered for Garsi-l, while the lowest values (2,17 g/L and 2,18 g/L) were obtained for Zaghavani 1. Juice extracted from pomegranates Garsi -1 was also characterized by the highest vitamin C content (0,250 g/L). Acid fresh juices could be wellconserved but were less appreciated. Their mixing with sweet juices considerably improved their properties and acceptability.

Anonymous (2006) reports the reason why fruit juice has recently grown rapidly is that people all over the world tend to buy natural and healthier products in order to live healthy and quality life. Companies have put consumers expectations and demands within the 1st rank among priorities. Consumers variety expectations sometimes focused on taste, sometime on healthier products. By keeping all those expectations in mind, pomegranate Mediterranean's mystic fruit, has intensively entered into the agenda of both producers and consumers. This is how Dimes pomegranate juice adventure started.

Khazaei and Kianmehr, (2006) experimented on seed extraction of pomegranate using a pneumatic methods. Here air under pressure is infused into the halves of the fruit so that the arils substantially disintegrate over a predetermined time period. For doing the tests, a prototype pneumatic robotic arm was designed and constructed. The system comprises a robotic arm controlled by a hardware system attached to a personal pc, fruit holder, air compressor and air nozzle through which compressed air is forced into the halved fruit. The robotic arm moves and controls the route of the air nozzle on the surface of the fruits. The route of the air nozzle and the time period (nozzle velocity) to blow out the air on the surface of the halved fruit was controlled by the hardware attached to the pc. In this research the effect of air pressure and route of the nozzle on the surface of the halved fruits, and number of passage of the nozzle were studied on percentage of removing arils from the pomegranate. Statistical analysis of the data showed that all the variables had a significant effect on removing the pomegranate arils ($p=0.01$). The best combination of the variable to extract the pomegranate arils was belong to air pressure of 8 bar, 4 times passage of nozzle and 8shape route. While best quality of the product was obtained by using the air pressure of 7 bar, 4 times passage of nozzle and 8shape route.

Shimlovich *et al.,* (2006) says that, major obstacle to the realization of the full potential of pomegranate is the difficulties involved in retrieving the internal edible seeds (arils). A novel method and a system was developed, which enables opening the fruit without cutting, extracting the arils with minimum damage, separating the arils from any extraneous materials and delivering clean arils to a packaging machine. A Beta site machine is currently under extensive testing and the results obtained so far indicate that it would be possible to extract 1.5 to 2 tons of arils a day in a one-lane machine, operated by 1-2 persons with an extraction efficacy of 95% and with 5% mechanical damage. The comprehensive fruit handling machine, moving the fruit through various stations to perform the operations of opening the fruit, exposing the opened fruits to the extracting mechanism, retrieving the extracted material and finally, separating the arils from extraneous material and delivering the clean and sorted arils to the packaging machine.

Hasan and Hasan, (2006) studied the effects of pressing on pomegranate juice quality. Fruits of three different cultivars were pressed at different pressures, time and shape as whole, pests and sacs only. Juice yield were increased up to 40% depending on pressure, time and cultivar but such juice was of excessive astringency and bitterness. It is mostly attributed to phenolic substances. These substances pass through juice by size reduction of the fruits before pressing. Thus best results in pressing were obtained by using whole fruit compared with juice sacs and size and size reduced pomegranates. Amounts of phenolic substances in fruit juice are increased by increasing pressure. It was evaluated that it wouldn't be feasible for both phenolic substance and pressing cost, if pressure is taken over 10 kg/cm^2.

Sagar, (2006) reported that the pomegranate powder could be prepared by steam blanching of the pomegranate seeds for 8 minutes, sulphured with burning of 10 g of sulphur powder1 kg of pomegranate fleshy seed for 4 hrs followed by drying in cabinet drier at a temperature of 58-60°C for 20 hrs and grinding the anardana into powder form by laboratory summit mixer grinder and sieving with 30 mesh sieve. The powder can be used with the milk and as churn form. Among the packaging material 260 g ALPE pouches were found to be best for the storage of pomegranate powder in respect of colour, flavor, texture, and other nutritional quality parameters, as compared to 200 g and 400 g LDPE, 200 g HDPE and 150 g PP (Polypropylene) pouches. For the storage of pomegranate powder in better quality and free flowing conditions the optimum RH 60% was found to be best when the critical and danger parts were at 7.78 and 6.82 percent moisture level respectively.

Panah, (2006) reveals that, harvesting and conservation of pomegranate is a main problem of producer in Iran, the production of pomegranate in spirited juice is a traditional solution. As some varieties have low level marketing, most of the medium quality fruits are used to produce inspirited juice. This opened different kinds of job and marketing opportunities. In this study, the traditional and industrial methods of pomegranate inspirited juice production and investigated. The proolem of pomegranate inspirited juice production is related to productivity and economic efficiency.

Hasnaovi *et al.,* (2006) studied pomegranate juice composition for 30 tunisian local cultivars. In total 6 anthocyanins, 8 organic acids and 6 sugars were analysed. Predominant sugars were fructose (7.21 to 10.61%) and glucose (5.67 to 8.31%). Traces of arabinose, sacchorose and galactose were detected but maltose and sorbitol were not detected. Among organic acids, the highest content was detected for citric acid (3.14%) in cultivar Mezzi 2, but average values showed that malic acid is the predominant one – Lower levels of succinic, oxalic, tartaric and ascorbic acids, traces of fumaric and no acetic acid were detected. Anthocyanin levels were quantified by comparison with an external standard of cyaniding 3 – rutinoside, based on the peak area and total anthocyanins content was calculated as a sum of individual pigments. The highest level (202.62 ppm) and the lowest one (11.48 ppm) were determined, respectively, in cultivars Mezzi 3 and Tounsic. In some cases, anthocyanin concentration seems to be positively correlated with red colour intensity of the juice.

Melgareja, *et al.,* (2006) says pomegranate fruits are rich in anthocyanins, pigements responsible for external and seed colour development. And the later is a crucial quality attribute for consumer acceptance. The goal of this study was to evaluate pomegranate jam anthocyanin content and colour development over time. Different temperatures and light regimes were tested during storage. Optimal storage conditions were achieved at 5°C and no light exposure at all.

Matapathi *et. al.,* (2006) revealed the procedure for wine making from pomegranate juice was developed and standardized. The three different cultivars of pomegranate comprising Ganesha, Arakta and Kesar are screened for their suitability for the wine production with three yeast strains viz., FWY-4, FWY-6 and Standard S. clipsoideus No.101. The wine produced from these cultivars was tested for its alcohol percent, residual sugar, tannins, aldehydes and esters. The overall acceptance of the wine was considered by organoleptic evaluation. Though all the three varieties of pomegranate were found suitable for wine making, the wine from Arakta variety scored maximum of 16 points and adjudged to be the best followed by wine from Ganesha and Kesar cultivars with standard wine yeast culture of saccharomyces elipsoideus No.101.

Khoshnam, *et al.*, (2006) said the physical characteristic, dimension, mass, volume, projected area and surface area are important factors in sizing systems. Mechanized filling of packs by numbering fruits is usually easier and cheaper than weighing packs. Market prepared fruits with the similar weight and uniform shape. Therefore grading fruit based on weight reduced packing and handling cost and also provides suitable packing patterns. The different grading systems required different fruit sizing based on particular parameter. Pomegranate mass was predicted by applying different physical characteristics with linear and non linear models as three different groups: 1) Single or multi variable regression of pomegranate dimension 2) Single or multi variable regression of pomegranate projected area (group 2) 3) Estimating pomegranate shape with geometric shapes based on volume (group 3), The 3rd, 2nd and 1st group model had highest correlation, respectively. The 2nd group models require electronic systems with digital camera for taking picture. The 1st group models except multi variable regressions, can grade as a simple mechanical system. The 3rd group model need complete electromechanical systems. Among systems grading pomegranate based on dimension (group 1), the best model is minor diameter with non linear relation M = 00.06 c – 4,11 c _ 143.56′ R2 = 0.91 with RSE = 15.43.

Majid and Javed, (2006) studied the determining effect of some parameters on extracting the pomegranate arils using a pneumatic method. In this research the effect of air pressure, nozzle diameter, route of the nozzle on the surface of halved fruits and number of passage of the nozzle on the surface of the fruits were studied on percentage of removing arils from the pomegranate. The experimental results revealed that best combination of the variables to extract the pomegranate arils was belong to air pressure of 7 bar, nozzle diameter of 3.3 mm and 4 times passage of nozzle on the surface of the fruit. For this combination of the variables, 100% of arils were removed. All the variables had a significant effect in removing arils ($p = 0.01$). The best route of the nozzle in this 8 shape route surface of the fruit was belong to traveling at 8 shape route.

Lan Xiang *et al* (2006) determined the ellalgc acid in pomegranate leaves, influence of varieties, seasons, processing methods in China. Pomegranate leaf has been used as herb tea in Shandong province in China for a long time with many pharmalogical effect. Presence of ellagic acid has aroused great interest to many scientist owing to its properties as anti carcinogenetic and antioxidant agents, the ellagic acid content in different varieties ranged from 0.13% to 0.646% dry weight, increasing significantly along with growing season, it accumulate former culture three months from June – August (0.1 – 0.4%) and reached the highest level in September and October with an exception of XHT variety highest at August (0.646%), while declining rapidly in September (0.594%) and October (0.478%). The most favorite cultivar DQPT in ZaoZhuang region contain ellagic acid of

0.169%, 0.359%, 0.397%, 0.431% and 0.525% during consecutive five seasons. The processing methods have different effect on the content of ellagic acid in pomegranate leaves. Steeping over one night or two night increased the amount of ellagic acid. Heating at 800 – 1000 had little effect of ellagic acid reaching from 0.37 to 0.891%, 1.138% and 1.767% respectively when leaves were heated on slow fire for 10, 20, 30 minutes.

Sachs *et al.,* (2006) reported on effect of modified atmosphere packaging on preservation of pomegranate fruits. The use of Xtend® Modified Atmosphere/ Modified Humidity (MA/MH) bulk packaging to alleviate these problems has become widespread amongst Mediterranean pomegranate growers and shippers in recent years, in particular since it is simple and cheep to use and does not require the use of expensive controlled atmosphere (CA) storage facilities. One major advantage of Xtend® MA/MH packaging in comparison to CA is associated with flexibility and the ability to remove a required amount of packaging upon demand without compromising the storage of the rest of the produce stored in the cold storage facility. After 16 weeks at 6°C followed by an additional 5 days at 20°C, weight loss was reduced from 7% in non-bagged control to 3.5% by packing in Xtend®. In the same trial, the extent and severity of scald was also considerably reduced in Xtend®. The incidence of scald fruit having 'moderate' to 'severe' browning after 16 weeks at 6°C was 2-11% in Xtend® packed fruit in comparison to 38% in the non-bagged control. Xtend® also reduced fruit and crown decay with an effect that was enhanced by treating with the fungicide, sportak before packing. The internal quality of the fruit was maintained in Xtend® after 16 weeks at 6°C as evidenced by the glossy appearance of the arils and white colour of the segments separating them. Taste tests conducted immediately after removal of the pomegranates from cold storage and five days after the fruit was removed from the bags and held at 20°C, revealed no impairment in taste. The ability of Xtend® MA/MH packaging to prolong storage of pomegranate for 16 weeks enables growers and shippers in essence to expand their marketing season.

Ram asrey (2006), says about post harvest management of pomegranate in India. Indian pomegranate is in good demand in Germany, England and Gulf countries. The peak availability months of pomegranate in India are January, February and March. The post harvest losses are very high and ranges between 25-30%. Harvesting is completed mostly manually. Sorted, graded and waxed fruits of A & B grades are packed in corrugated fiber board (CFB) boxes and sent in upcountry market of metropolis through railwaygons and non-refrigerated trucks. More than 98% produce is utilized as fresh for table purpose and rest is in processing industry. Most of the Indian varieties give juice recovery 42-43% on whole fruit basis. Carbonated drinks, wine, syrup and Anardana are major processed produce of pomegranate.

Moghadam and Nikkhah(2006), recommended that, heat treatment is one of the method for controlling the storage pest of fruits. Different methods of heat treatment comprise hot steam, hot air and hot water. In order to increase shelf life and keeping quality of pomegranate fruit this study was conducted during 1999. The treatments were heat treatment (hot water 55 and 65 degrees centigrade for 2 minutes and hot aroid 45 and 51 degrees centigrade for 25 minutes) packaging (shrink wrap and without wrap) and storage (time 0, 30, 60, 90 and 120 days). A split in time factorial experiment was laid out adopted completely randomized design with 4 replications. All data were subjected to analysis of variance and Duncans range tests were used to compare the treatment means. Results showed that, heating the fruits in water at 55°C for 2 minutes in comparison with other treatments caused the least decay in packaged and stored fruits also during the storage time, weight loss and decay increased in fruits. Besides weight loss of fruits in shrink, wrap correct for fruits that preserved in cold storage or normal storage after heat treatment for 4 months.

Palou, and del Rio, (2006), studied the assessment of fungal pathogen causing post harvest decay of pomegranate in south Spain. Spain is leading European Union producer and exporter of pomegranate. 90% of the area is located in Alicante and Murcia province, at present there is a tendency to increase storage shelf life in off season markets, chilling injury, weight loss and decay are the most important problems limiting storability of pomegranate. Commercially grown pomegranate cv "Mollar de Elche" from two different orchards were used in first season study to assess both latest pathogen (field infection) and wound pathogen (post harvest infections). Healthy pomegranate were either surface disinfected with 0.5% sodium hypochlorite or artificially wounded in the rind and placed in humid chamber at 20°C for 2 months. Decay was periodically assessed on intact pomegranate stored at 5°C for upto 6 months. Isolates of pathogen causing fruit decay were plated in potato dextrose agar (PDA) Petri dishes and incubated at 25°C for further macroscopic identification. Although identification of few pathogen causing minor decay is pending, preliminary results indicated that the main casual agents of latest and wound infection were *Botrytis cinerea and penicillium sp.* respectively and the same pathogen were frequently isolated from cold stored fruits, another pathogen at 20°C observed was *Aspergillus niger.*

Inci cinar and Levent (2006) reported storage stability of seed powder and black berry added to pomegranate juice, in Turkey. Pomegranate in Kahraman Maras is grown for its sweet or acidic taste, its use has significantly increased (as juice or concentrate) owing to its well recognized health benefits. Pomegranate is not harvested year round, thus processing to extend the shelf-life of its products is inevitable. Processing does inevitable determine however considerable colour quality loss. Color play an important role in visual acceptance of fruit juice by consumer and degree of color loss is hence important to assess the ultimate quality of juice. The addition of highly acidic (7.8 – 15.4%) seed powder or dark colored anthocyanin rich black berry help improving colour stability during storage information on color degradation kinetics along with other parameters is used to predict storage stability of juice. In this study sweet type pomegranate from Kahraman Maras were used. Pomegranate seed power (1, 5, 10 % w/w) and black berry juice (10, 20, 30, 40, 50% v/v) were then added to their juice (15% soluble solids) and pasteurized and stored at 25 and 4°C. Anthocyanin index and browning index values were determined spectrophotometerically and the effects of either seed powder or black berry addition and storage condition on the juice stability. The reaction rate constant and activation energy were calculated using Arrhenius type relation of measured colour and storage time.

Zolfeghari *et al.,* (2006) reported reducing chilling injury and decay incidence in pomegranate fruit in Masas save by methyl salicylate. Pomegranate fruit manifest chilling injury when exposed to low, non-freezing temperature than can be the most limiting factor to storage fruits in low temperature for long time. There are different concentration of Methyl salicylate (0, 8, 16, 24 μl. L^{-1}) in inducing chilling tolerance and decay incidence of pomegranate fruits (Malas Save) were investigated. Qualitative attributes of fruits was investigated during low temperature storage and also after transferring that to high temperature as or shelf life. Response of treated fruits to low temperature storage demonstrated that there is a significant difference in comparison with untreated control fruits. The pomegranate fruits treated with 24μl. L^{-1} Methyl salicylate had the minimum chilling injury and decay incidence at the end of storage at 2°C. The results indicate Methyl salicylate suppress chilling injury and water loss and prevented external appearance in pomegranate fruits without detrimental effects on internal fruits quality. Further there was no significant difference was found between treated and control fruits for total soluble solids, titrable acidity, pH.

Moghadam and Nikkah (2006) studied the effect of post harvest heat treatment in reducing the pomegranate moth damage. To increase awareness of consumers that many chemical treatments of fruits and vegetable for pest and disease control are harmful for human, there is a need for a physical method for disinfection against pest and disease and one of them is temperature. This study was conducted in 1998 to control fruit moth damage and maintaining fruit quality. In the first phase fruits were dipped in hot water (25, 50, 55, 60, 65, 70, 80 °C) for 1-2 minutes and in the second phase fruits treated with hot air (45, 47, 49, 51 °C) for (15, 20, 30, 40, 45 minutes). After heat treatment fruits packed (without wrap and shrink wrapping) in each phase. The results showed that (55, 65 centigrade degree water for 2 minutes 45, 51 centigrade degree air for 25 minutes significantly affect on reducing pomegranate moth damage.

Singh *et al.*, (2006) says that, the conventional utilization of wild pomegranate fruit is by drying the seeds along with pulp (arils), which constitute the product 'anardana'. The dehydrated arils are acidic and help in improving mouth feel and digestion. Anardana finds wide application in traditional Asian medicine for several disorders, such as stomachache, diarrhea and bronchitis. The drying temperature is optimized in this study for better quality retention during storage. The arils were dehydrated at 50, 55, 50 °C for 7 hours retained desirable and acceptable quality parameters (titrable acidity 7.8%, vitamin C as ascorbic acid, 15.2 mg/ 100 g) up to 180 days of storage. The red colour (Hunter 'a' value) of the product changed with storage duration. The product obtained by drying at 55°C had permissible enzymatic browning and microbial load within acceptable limits.

Aquino *et al.*, (2006) revealed the efficacy of post harvest dip treatments, with fludioxonil (FLU) of fruits of 'primosole' pomegranate in decay control of artificially wounded fruits was investigated over 3 weeks of storage at 20°C, 90% relative humidity, while residual levels of FLU were detected at time 0 and after 2 weeks of storage. The incidence of decay in control fruits was approximately 60 and 100% after 1 and 2 weeks of storage, respectively. Hot water dipping for 2 minutes at 50°C resulted in about 45% reduction of decay after 1 week of storage with respect to untreated fruits, but after 2 weeks of storage 100% of fruits decayed. FLU treatments were very effective in controlling decay, caused by *Botrytis spp. and penicillium spp.,* during the first two weeks of storage, especially when applied at 150 mg/L at 50°C. but was less effective against heart rot, caused presumably by *Aspergillus* niger and *Alternaria* spp., Total residues were recovered from the skin of pomegranate, while the presence of FLU in the edible part was not detectable. Residue levels in fruits were affected either by FLU concentration and dip temperature. Following treatment FLU applied at 20°C at the rate of 100 mg/liter produced a residue of 0.25 mg 1 kg (active ingredient, whole fruit basis) this value increased respectively by 2.7 and 4.9 fold when FLU was applied at 25, 75 or 150 mg/liter at 50°C residue level were 0.22, 0.38 and 1.26 mg/kg. A reduction of about 30% of residues occurred after 2 weeks of storage only in fruit treated with 150 mg/L at 50°C or in those treated with 600 mg/L at 20°C of FLU while little changes took place in the other treatment.

Ranjbar *et al.*, (2006), conducted experiment during August 2004 to evaluate the effects of pre harvest calcium chloride and post harvest hot water treatment as well as the use of polyethylene bags for packaging on the quality of pomegranate fruit during cold storage. For this purpose calcium chloride sprayed at 3 concentrations 1%, 2% and control). Pomegranate fruits of Malas Torsh cv. were harvested at fully ripe stage and treated with hot water 45, 50 and 55°C for 20 minutes, 1 minute and 30 second). Thereafter each fruit was warped in a polyethylene bag and stored in cold storage room at 1°C and relative humidity of

85% for 3 months. The experiments were conducted using a completely randomized design of three replications. Polyethylene as well as marketability of fruits during storage, while showing no effect on pH, TSS and fruit dry matter. Also it was observed that fruits treated in 50°C hot water showed significant improvement of quality while treatment with 45°C hot water failed to show such effects. In total, a combination of calcium chloride and polyethylene bag packaging along with 50°C hot water treatment caused improvement in general quality of stored fruit and decreased fungal infection and chilling index.

Bayram and Dundar (2006) conducted a study to prolong the storage life of pomegranates grown in Adana, Turkey by using modified atmosphere packaging and shrink film wrapping with polyolefin film. In this study, the fruit was harvested at optimal harvest time and divided into three groups and placed in plastic boxes. The first group of the fruit was covered by ordinary PE plastics and sealed. The second group was covered by polyolefin film. The control was kept uncovered during the experiment. All the fruit samples were stored at 6°C temperature with 90% RH. During the storage fruits evaluated every 30 days for weight loss, titrable acidity, total soluble solids content, skin colour, pH and anthocyanin concentration. Further more, fungal and physiological decays and the changes of CO_2 and O_2 concentration of the fruits were also recorded.

Rouhani and Hadjighasemi, (2006) studied the comparative activity of different colorant fractions separated from two kinds of Iranian pomegranate rind. In their experiment, different colorant fraction of two kinds of Iranian pomegranate rind (custom and black rind) were extracted through methanol, total phenolic compounds, radical scavenging capacity (RSC) and antimicrobial activity were determined in each extract. RSC (%) was measured by DPPH method and the results were compared with BHT standard solution. All of fractions exhibited high antioxidant activity. The custom pomegranate rind extracts showed higher activity than the black. Rind extracts of two species possess strong in vitro antimicrobial activity against the colonies of the bacteria tested as *Bacillus, Staphyloccocus aureus, Staphylococcus epidermidis, Enterococcus faecalis, klebsiella pneumoniae, Escherichia coli, pseudomonas aeruginosa.*

Rouhani *et al.*, (2006) found that synthetic food colorants are steadily replaced by natural ones because of consumer preferences, Anthocyanins are natural, water soluble, non toxic pigments displaying a variety of colours from orange to blue. Because of their antioxidant properties, they may also have a beneficial influence on human health. Anthocyanin pigments are rich in pomegranate. A limiting factor for the incorporation of Anthocyanin is the low stability of these pigments, which was influenced by several factors such as pH, temperature and different agents. The statistical methods are widely used by quality engineering scientists to compare the effect of multiple variables simultaneously. Taguchi experimental design was employed to test the factors effective on extraction process of coloured extraction of two types of Iranian pomegranate (custom and black rind). The optimum condition for extraction process was obtained as: soaking time, 48 hr, type of acid, HCL, acid percentage 5%, organic solvent, methanol, organic solvent percentage 10%. They found that best conditions for storing are -4°C, 5% tannin as stabilizer, and pH =2.

Palma *et al.,*(2006) they have evaluated the chemical and organoleptic characteristics of minimally processed seeds of pomegranate 'primosole' after packaging in polypropylene trays (150 g each) sealing with a 40 μm thick polypropylene film and subsequent storage at 5°C for 10 days. The chemical analysis of centrifuged and filtered juice revealed no significant changes after 10 days of storage. Titrable acidity increased slightly (from 0.27% to 0.30%) while a little reduction occurred in total phenolic content, which decreased from 1492 to 1339 mg/L. Relevant changes were recorded in juice colour, with a significant increases in a* values, decreases in h° angle, while changes in L*, b* and chroma parameters were not significant. By the end of storage, pomegranate seeds did not exhibit visible symptoms of decay and no undesirable effect was recorded on quality traits such as the development of off-flavour and off-taste.

Ghasemnegad, *et al.,* (2006), says that, susceptibility to chilling injury in pomegranate fruits is the main limiting factor for storage of fruits at low temperature. Inducing chilling tolerance make storage of fruits possible even at low temperature for a long time. In this study, effect of different concentrations of methyl jasmonate (8, 16 and 14 μl.L) on inducing chilling tolerance of pomegranate fruits (Malas Save) was investigated. Qualitative attributes of treated fruits was studied during low temperature storage and also after transferring to high temperature to test the shelf life. The results showed that methyl jasmonate suppressed chilling injury and water loss and preserved external appearance in pomegranate fruits without any abnormal effects on internal fruit quality. In general, with increasing methyl jasmonate concentration, chilling injury was reduced significantly in fruits. The highest chilling injury index was found in control fruits and the lowest one was for 24 μl, L during storage at low temperature and also after placing at higher temperatures. Furthermore, no significant differences were found between treated and control fruits for total soluble solids (TSS) and titrable acidity content.

Yahya Selahvarzi *et al.,* (2009) carried out the Bio-index study in pomegranate chilling injury. Unexpected chilling and Freezing of winter damaged pomegranate orchard in Ferdows region of Iran. Using reliable method to estimate chilling injury would help gardeners to select the most tolerant cultivars and also file insurance claims. Therefore three chilling estimation methods i.e., visual observation, regrowth and electrolyte leakage were compared in a factorial experiment based on complete randomized design. Three pomegranate cultivars from Ferdows i.e., Shisha Kabe, Malas Torsh and Shalgami were selected for the study after chilling injury of -25°C (2008). According to results, not significant difference was observed among cultivars, estimation methods and their interactions, it was concluded that although regrowth method (used as control) was the most precise and obvious one, the fastest results could be tracked by electrolyte leakage method. Visual observation that is based on anatomic method (tracking cortex and cambium browning) should less accordance with the other methods probably because of diagnosis errors.

Sarig and Galili, (2009), Pomegranate is an ancient fruit crop, which has been around for more than 3000 years, inspite of its long history, it remained for many years as a minor crop with relatively limited marketability. The major obstacle to the realization of the full potential of this unique fruit was the difficulties involved in retrieving the internal edible seeds (arils). However, in recent years major developments in pomegranate post harvest technologies, coupled with the revelation of the unique medical benefits of the fruit, led to a dramatic increase in acreage world wide and hence a significant increase in its production. The medicinal benefits of the pomegranate have already been well known in the traditional medicine of many countries. However, only in recent years well accredited medical studies have been carried out in several countries, but primarily in Israel, have proven that pomegranate juice can reduce blood pressure, reduce LDL oxidation and reduce the atherosclerosis. Moreover, it has shown major promise to combat prostate and breast cancer. A novel and unique method and a system were developed at the Agricultural Research Organization in Israel, which enables the retrieval of clean, dry and damage free arils, which are packed in MAP packages to ensure storing on chilled shelf-life for up to 12 days. The machine is currently being commercially manufactured under license by Juran Metal works' in Israel and is being offered as post of a processing system of varied configuration from one lane configuration with a throughput of 200 kg/h of clean and ready to be packed arils, to a multi-head machine, which can reach a capacity of more than 800 kg/h. Likewise, different configurations according to the end use fresh packed arils, or processed arils for juice and/or wine. A more recent development is the process of freezing the arils, developed by Israeli scientist in collaboration with "Juran Metal Works". The process, which is still not fully commercialized, complements the arils extraction process by offering both, flexibility and optional lining for the product consumption. Another important development in the acquired capability to store pomegranate in cold storage up to 5 months while the aforementioned developments gave a tremendous boost to the pomegranate industry, several issues still remain unsolved and present a challenge for future development. A non-destructive internal quality mechanism should be developed beyond the current existing prototype and integrated into the extended to ensure a safe handling for distance markets. Concurrently, different process should be developed to process the various parts of the fruit juice, arils, seed and rind for new beneficial food supplements and cosmetic products.

Singh *et al.,* (2009), reported the effect of minimal processing of pomegranate arils on quality maintenance and shelf life. They says that consumption of pomegranate mainly as fresh fruit, but the difficulty of peeling limits its consumption. The commercialization of 'ready-to-eat' minimally processed fresh arils could be a good alternative for the national and international market. The minimal processing consists of washing with sanitizing agents to reduce the initial microbial load, pH modifications use of antioxidant agents, temperature control and other, to control partially the high perishability of the fruits. On the other hand, the use of polymetric unfavorable conditions for the action of many contains microorganisms. During post harvest handling, strawberry fruits show browning produced by oxidation of phenolic compounds indicating that the stabilization of anthocynic pigments is essential in order to achieve a good quality. Experiment was taken up to study the effect of sanitizing agents, antioxidant mixture solution and packaging with semi permeable films on the quality maintenance and shelf life of minimally processed pomegranate arils of variety Mridula stored at 5 ± 1°C for 16 days, pomegranate arils treated with chlorinated water for 5 minutes (sodium hypochlorite 100 ppm) + Antioxidants for 30 seconds (5% w/v citric acid and Ascorbic acid) and packed in 25 inch semi permeable PE film and stored at 5 ± 1°C maintained quality, colour and shelf life up to 16 days with good physical and quality conditions, for their commercialization. Anti-oxidants were positive to reduce spoilage and retain colour, quality and chemical composition of arils.

Gawade *et al.,* (2009), studied effect of intermittent drying technique on quality of anardana from pomegranate arils (Cv. Ganesh). The effect of pretreatment, drying time and drying cycles (intermittent), on drying characteristics and quality of anardana was studied. The pretreatment of citric acid with 2,3 and 4 percent as given prior to drying. The drying of arils was carried out at 50 and 60°C temperatures. Intermittent drying cycles viz., 1:0.5, 1.5:05 and 1.5:1 hour were used during drying. The pomegranate arils were dried from initial 400 percent moisture content (d.b) to final 6.75 – 7.50 percent moisture content (dry weight basis) of Anardana. The aril without pretreatments was taken as control. The chemical properties of anardana were also studied. Total drying time varied between 12 to 5.5 hr (with tempering) whereas effective drying time ranged between 3 to 6 hr (without tempering). The maximum drying rate was obtained in drying while minimum drying rate was obtained in tempering in falling rate period. The max drying rate was found to be 293.01 to 318.88 g/hr – 100 g of dry matter during drying cycle 1:1 hr at 50°C. Maximum drying rate of 157.16 to 168.97 g/ hr – 100 g of dry matter was recorded in intermittent drying cycle 1:1 hr at 50°C.

Generally 0.44 to 9.82 g/hr – 100 g of dry matter drying rate was observed during tempering period. The drying curve obtained was exponential in nature. The T.S.S and acidity of anardana was found in the range of 32.16 to 36.98 percent and 1.18 to 5.28 percent, respectively because of citric acid concentration. Slight variation in sugar content was due to inversion of non reducing sugar to reducing sugar in presence of citric acid and temperature. The reducing sugar, non reducing sugar and total sugar of anardana varied between 12.72 to 16.21 percent, 1.86 to 2.27 percent and 14.87 to 18.14 percent, respectively. The maximum acceptability of anardana was found in the treatment of 3 percent citric acid dried at 60°C for intermittent drying time of 1.5 to 0.5 hr.

Amos Levin (2009), reported the impact of practical application of cold storage of pomegranates in a controlled atmosphere in Israel. Approximately 2000 hectares of pomegranates are planted in Israel in various growing regions between elevations of 100 to 700 meters above sea level. Considering that within two years expected yields will reach approximately 60,000 tons, the question arises of post harvest treatment. The main variety is wonderful which is suited to lengthy storage periods of more than three months, thus allowing an extension of marketing season. Israel has advanced controlled atmosphere cold storage facilities for apples, peers and kiwi with a capacity of 100,000 tons. It can be adapted to pomegranate storage wonderful variety of pomegranate fruits reach a minimum average size of 400 grams each. Ripening in October, their internal and external colour is an intense red. In the European markets, maximum prices are obtained at Christmas and New Year and in the months of February and March as Easter approaches and quantities imported in to Europe decline. Key limiting factors encountered while attempting to extend the cold storage period i.e.,

- Development of scald: raising the CO_2 concentration reduces scald and rots that appear in storage. Storage in 2% oxygen was also found to reduce scald.
- Fungal rots caused by *Aspergillus niger*, *Penecillium sp.* and *Botrytis cinerea* : research indicates that a high concentration of CO_2 causes damage to the peel and internally in wonderful. Concentrating of 3-6 percent of combination with 2% oxygen improves fruit quality reducing scald, chilling and storage rot. Shrinkage due to water loss, expressed as weight loss under conditions of controlled atmosphere it is usual to wrap the fruit containers in polyethylene sheets to prevent weight loss due to water loss by the fruit.
- Chilling injury that causes browning of the membranes surround the arils : chilling damage develops in stored fruit at temperatures below 5°C. Storing pomegranates in 2% oxygen reduces chilling damage.
- Molding of the stamens in the blossom end of the pomegranate, fading of the colour of the arils and development of an off-taste in storage which damage fruit quality.

Generally storing temperature ranges between 6-8 degree centigrade and Relative Humidity of 90%. In summer when all necessary measures are employed and most reliable knowledge is applied. Pomegranate can be stored for 4 months and the fruits removed at the end of the period will be suitable for marketing in Israel's export markets.

Mahadeviah (2009) developed the technology of value added products in pomegranate, which is known for medicinal and therapeutic values. Pomegranate is seasonal and available only for few months in a year. Hence it is not possible to consume fruits through out the year. The value added products are Juice, RTS (Ready to serve), Squash, Jelly/bar, toffee, Anardana.

Shivlal *et al.,* (2009) studied the effect of coating materials on shelf life of pomegranate fruit under ambient storage condition. Pomegranate rich in vitamin A, B and minerals. In temperate region harvested in September – October, since the fruits are having short shelf life under ambient storage condition which causes heavy monetary losses, the only option is to store in cold storage or by coating with various coating material. In the present study Dholka variety shelf life was enhanced by combination of calcium chloride, paraffin wax, carbendizim coating { paraffin wax 4%, Ca (2%), Ca (4%), Ca (2%) + paraffin wax (4%), Ca (4%) + Ca (4%) + paraffin wax (4%), Ca (2%) + paraffin wax (4%) + carbendizim (0.1%)}. General appearance i.e., colour percentage weight loss and organoleptic parameter were studied. Maximum retention of fruit colour. Maximum retention of fruit shape and minimum weight loss up to 75 days was recorded in treatment Ca (4%) + paraffin wax (4%) + carbendizim (0.1%). Highest soluble solids were in Ca (4%), maximum acidity was (0.1%). However maximum total and reducing sugars were found in Ca (2%) + paraffin wax (4%), while highest ascorbic acid content was recorded in paraffin wax (4%) at 75 days after storage.

Thakur *et al.*, (2009), reported development of tool and machine for safe separation of arils from pomegranate. The tool and machine which is capable to break the pomegranate fruits in a manner of two irregular halves, separates arils, about 35-40% during breaking process and makes the remaining arils loosen that allow easy separation further by finger tips. The tool is designed considering its appropriate functioning to accommodate medium size of fruit which is distributed in maximum number on a particular plant. The machine namely "CIPHET – pomegranate Aril Extractor" relates to a mechanical and continuous system for processing of whole pomegranate of any size, shape and variety; and for recovery of clean, whole and undamaged arils. The machine is capable to process the whole pomegranate at a rate of approximately 5.0 q per h with an extraction capacity of 92-94% and with little damage of arils that is about 2-3%. The tool and machine was tested for five popular varieties (Ganesh, Mridula, Muscat, G-137, Sinduri) of pomegranate and the performance was evaluated in terms of arils in comparison to the arils separated by conventional method using knife and fingers. The performance of tool and machine was found encouraging as it makes the task of safe separation of arils from the whole fruit effortless.

Rokhade *et. al.*, 2009 says that, the seeds along with fleshy portion are dried and commercially marked as anardana which is widely used as condiment and as an acidulant in Indian curries chutneys etc in North India. Production of anardana is one of the important methods of value addition. However, there is large variation in the method of production and sensory quality of the product. The possibility of utilizing low quality pomegranate fruits for preparation of anardana was investigated. The arils obtained from both good low quality fruits were subjected to a pretreatment of sodium benzoate and dried under either open sun or in a solar drier. Organoleptically acceptable quality anardana could be prepared, after pretreating the arils with 600 ppm sodium benzoate for 10 minutes and drying in solar drier. The sensory evaluation revealed that the anardana prepared from low quality fruits were on par with that prepared from good quality fruits with a score of 4.60 (out of 5.0) for colour and appearance, 4.63 for texture, 4.61 for flavour, 4.59 for taste and 4.3 for overall acceptability.

Dheeraj Singh *et. al.,* (2009) indicated a experimental evidence regarding drying characteristics and organoleptic quality suggested that optimum tray load for efficient and economical drying was 1.23 kg/m^2. Pomegranate seeds blanched for 5 minutes in boiling water gave the best quality anardana having bright maroon colour under sun drying as well as under cabinet drying conditions. Mean values for moisture differed significantly in sun dried (15.72%) and cabinet dried (9.33%) samples. Anardana samples had 34% more acidity in cabinet dried product than sun dried product irrespective of pretreatment used. Non-enzymatic browning was maximum in cabinet dried pre-treated samples (steam blanched along with sulphuring). Similarly anthocyanin content differed significantly among different samples and was maximum (0.907) in steam-blanched cabinet dried samples and minimum in sun dried control samples. In organoleptic evaluation also overall mean score for cabinet dried samples and minimum in sun-dried samples was significantly higher (7.42) than sun dried samples (6.27). With regard to pre-treatments, no statistically significant difference was observed for overall organoleptic quality of anardana samples. In general drying was faster and quality of anardana was better in a cabinet dryer as compared to sun drying.

Singh *et. al.,* (2009), revealed that pomegranate is being cultivated in large area in arid and semi and tropics, where high temperature and low humidity prevail. Storage and packaging become very important aspects to maintain the quality of harvested fruits. During storage, formation of black colour of arils and fungal attacks are important factors for consideration under arid and semi arid conditions fruits need proper post harvest handling including packaging and suitable storage.

Keeping these problems in view, efforts were made to study the effects of different post harvest treatments, packaging and storage of pomegranate on its shelf life and quality. The shelf life of pomegranate could be extended up to 90 days along with quality (PLW, TSS, acidity, shape, colour, vitamin C) and shape, when fruits treated with 4% calcium chloride, packed in 5 ply corrugated fibre board (CFB) boxes and stored at low temperature (5±1°C) having RH 85 percent.

Kshirsagar *et. al.,* (2009) experimented a technology of preparation of carbonated beverage from pomegranate fruits (Cv. Ganesh), to evaluate chemical and sensory properties of the carbonated beverage, to observe the changes in composition and quality of carbonated beverage during storage and also to work out the economics of the carbonated beverage of pomegranate juice. The yield on whole fruit basis and aril weight basis was 43 and 68 per cent respectively. The carbonated beverage was prepared with 10, 15 and 20% pomegranate juice with blending of ginger juice at 1, 2 and 3 percent by maintaining the Tss at 15°B and acidity to 0.30 percent. Based on organoleptic evaluation carbonated beverage with 10 per cent pomegranate juice and 1 percent ginger juice was best among all the levels. During the storage, Tss, pH, reducing sugars increased while titrable acidity, ascorbic acid and anthocyanin contents decreased. The total cost of production of carbonated beverage from pomegranate juice was Rs.9.41 per litre.

Rokhade *et. al.*, (2009), investigated the possibility of utilizing pomegranate juice for preparing carbonated beverage. Pomegranate beverage prepared by using different proportions of fruit juice, sugar and citric acid were carbonated using 60, 80, 100, Psi carbon dioxide gas and stored. Good quality carbonated RTS pomegranate beverage could be prepared with recipe containing 100% fruit juice to 5% citric acid adjusted to 65% TSS by addition of sugar and carbonated with 80 Psi carbon dioxide gas which could be stored up to 60 days. This carbonated beverage had highest organoleptic scores (out of 5.0) for colour and appearance (3.32), taste (4.08), flavour (3.18) and overall acceptability (4.02) compared to other treatments.

More *et. al.*, (2009) evaluated the chemical and sensory properties of the carbonated beverage, to observe the changes in the composition and quality of the carbonated beverage during storage and also to work out the economics of the carbonated beverage of pomegranate juice. The yield on whole fruit basis and aril weight basis was 42 and 63.5 per cent, respectively. The carbonated beverage was prepared with 10, 15 and 20 per cent pomegranate juice with blending of ginger juice at 1, 2 and 3 per cent by maintaining the TSS at 15°B and acidity 0.30 percent. Based on the organoleptic evaluation carbonated beverage with 10 percent pomegranate juice and 1 per cent ginger juice was best among all the levels. During storage, TSS, pH, reducing sugars, non-reducing sugars and total sugars increased while titrable acidity, ascorbic acid and anthocyanin contents decreased. The total cost of production of carbonated beverage from pomegranate juice was Rs.9.41 per litre.

Rokhade et al., 2009 says that, the seeds along with fleshy portion are dried and commercially marked as anardana which is widely used as condiment and as an acidulant in Indian curries chutneys etc in North India. Production of anardana is one of the important methods of value addition. However, there is large variation in the method of production and sensory quality of the product. The possibility of utilizing low quality pomegranate fruits for preparation of anardana was investigated. The arils obtained from both good low quality fruits were subjected to a pretreatment of sodium benzoate and dried under either open sun or in a solar drier. Organoleptically acceptable quality anardana could be prepared, after pretreating the arils with 600 ppm sodium benzoate for 10 minutes and drying in solar drier. The sensory evaluation revealed that the anardana prepared from low quality fruits were on par with that prepared from good quality fruits with a score of 4.60 (out of 5.0) for colour and appearance, 4.63 for texture, 4.61 for flavour, 4.59 for taste and 4.3 for overall acceptability.

Dheeraj Singh *et. al.*, (2009) indicated a experimental evidence regarding drying characteristics and organoleptic quality suggested that optimum tray load for efficient and economical drying was 1.23 kg/m^2. Pomegranate seeds blanched for 5 minutes in boiling water gave the best quality anardana having bright maroon colour under sun drying as well as under cabinet drying conditions. Mean values for moisture differed significantly in sun dried (15.72%) and cabinet dried (9.33%) samples. Anardana samples had 34% more acidity in cabinet dried product than sun dried product irrespective of pretreatment used. Non-enzymatic browning was maximum in cabinet dried pre-treated samples (steam blanched along with sulphuring). Similarly anthocyanin content differed significantly among different samples and was maximum (0.907) in steam-blanched cabinet dried samples and minimum in sun dried control samples. In organoleptic evaluation also overall mean score for cabinet dried samples and minimum in sun-dried samples was significantly higher (7.42) than sun dried samples (6.27). With regard to pre-treatments, no statistically significant difference was observed for overall organoleptic quality of anardana samples. In general drying was faster and quality of anardana was better in a cabinet dryer as compared to sun drying.

Kad *et. al.*, (2009), observed that effect of various extraction methods and clarifying agents on quality, preservation and storage of pomegranate juice. Four extraction methods viz., muslin cloth, screw juice extractor, basket press and house hold mixer for four clarifying agents viz., gelatin, polyvinyl pyrrolidone (PUP), bentonite and pectinase was used to get better quality (clear) juice. The clarified juice (using 0.05 percent (PUP) was further pasteurized at 60, 70 and 80^{0}C for 30 minutes and potassium metabisulphite (KMS) (50, 100 and 150 ppm) was added as a preservative. The effect of pasteurization temperature and preservative concentration (ppm) on storage was studied. The juice was stored at room temperature 27±2°C and low temperature storage (5±1°C). The pasteurization temperature and preservative concentration (ppm) was optimized. The juice yield of 61, 55, 48 and 40 per cent was obtained in case of house hold mixer, screw juice extractor, basket press and muslin cloth, respectively. The quality of juice was found better using muslin cloth method and 0.05 percent polyvinyl pyrrolidone. The effect of pasteurization temperature and preservative concentration on shelf life of juice was found significant. The juice pasteurized at 70°C for 30 minutes and treated with 100 ppm KMS could be stored for 120 days at room temperature and 270 days at low temperature storage. The TSS, acidity, ascorbic acid and total sugars of this juice was found 16.50, 0.47 – 0.48, 0.010 (10 g/100 ml) and 14.00 – 14.10 per cent, respectively after the storage for 270 days.

Masalkar, *et. al.,* (2009) conducted a study to elicit the effects of eleven post harvest treatments and two storage conditions viz., room temperature and cool store on shelf life of pomegranate fruits. Their effects were assessed by factorial completely randomized design. The treated fruits of pomegranate were stored at room temperature (9.50 to 35.00°C; 22 to 93% RH) and cool store (8°C; 85 to 95% RH). There was decrease in juice content, acidity while increase in TSS, reducing sugars, total sugars, specific gravity in both the storage conditions. In general, the cool store temperature was found to be better than that of room temperature as far as shelf life and physico-chemical qualities of fruits. Post harvest application of waxol (12%) + $CaCl_2$ (4,6, or 7%) + carbendizim (0.1%) to pomegranate fruits results in better retention of sensory quality attributes during storage. The shelf life of untreated fruits was found to be 16 days at room temperature. However, the fruits could be stored up to 35 days when treated with waxol (12%) + $CaCl_2$ (4%) + carbendizim (0.1%) at room temperature. The shelf life of untreated pomegranate fruits was 68 days. However, the fruits treated with waxol (12%), $CaCl_2$ (2%), $CaCl_2$ (4%), $CaCl_2$ (6%), waxol 12% + $CaCl_2$ (2%), waxol (12%) + $CaCl_2$ (4%), waxol (12%) + $CaCl_2$ (6%), waxol (12%) + $CaCl_2$ (2%), + carbendizim (0.1%), waxol (12%) + $CaCl_2$ (4%) + carbendizim (0.1%) and waxol (12%) + $CaCl_2$ (6%) + carbendizim (0.1%) the shelf life was extended up to 75, 92, 94, 94, 95, 96, 97, 97, 102 and 100 days respectively when stored in cool store. The pomegranate fruits (Cv. Ganesh) should be stored in cool store at 8°C; 85 to 95% RH with dippling treatments of waxol (12%) + $CaCl_2$ (4, 5 or 2%) + carbendizim (0.1%) coupled with corrugated fibre board box packaging in order to improve their shelf life for pretty long period, which will help in marketing of pomegranate fruits around the world.

Internal breakdown of fruits

Unfortunately, the production of good quality fruits of many verities is be set by development of serious malady known as internal breakdown or internal blackening, wherein, externally the fruit appears quite normal but its arils become soft having light creamy brown to dark, blackish brown colour. The incidence of internal breakdown crops up after 150 days of anthesis and its intensity increases if the fruits are left on the tree up to 165 days. The incidence of browning increases with increase in weight of fruit. It is reported that TSS, acidity, ascorbic acid, total sugars, reducing sugars, calcium, phosphorus and the enzyme catalase were low whereas non-reducing sugars, starch, tannins, nitrogen, potassium, magnesium, boron, polyphone oxidize and peroxides enzymes were high in affected arils than the healthy ones.

Table 1: Production cost of homemade Pomegranate Wine

Sl. No.	Items of wine making	Cost/quantity
1	Cost of 1 kg fruits (Rs)	10.00
2	Yield of juice (ml)/kg fruit	500.00
3	Cost of added chemical (Rs)	2.00
4	Processing cost (1/3rd of raw material cost, (Rs))	4.00
5	Wine yield from 1 kg fruit (ml)	400.00
6	Cost of one litre wine (Rs)	40.00

Table 2: Organoleptic Evaluation of homemade Pomegranate Wine

Sl. No.	Quality parameters	Dry wine	Sweet wine
1	Appearance (2)	2.00	2.00
2	Colour (3)	2.50	2.50
3	Aroma (2)	2.00	2.00
4.	Bouquet (1)	1.00	1.00
5	Vinegar (2)	1.50	1.75
6	Total acidity (2)	1.50	1.75
7	Sweetness (1)	0.00	0.50
8	Body (1)	0.50	1.00
9	Flavour (2)	1.00	1.50
10	Astringency (2)	1.00	1.00
11	General quality (2)	2.00	1.00
12	Overall acceptability (20)	2.00	1.00

(Figures in the paranthesis are maximum allotted marks for each parameter).

Table 3: Organoleptic evaluation of pomegranate sweet wines (mean average of three judges)

Wine	Spices added	Scores given by panel of judges in 20 point scale.											
		Appearance (2)	Colour (3)	Aroma (2)	Boaquet (1)	Vinegar (2)	Total acidity (2)	Sweetness (1)	Body (1)	Flavour (2)	Astringency (2)	General quality (2)	Overall acceptability (20 points)
Ganesha	Clove	0.50	1.00	1.00	1.00	2.00	1.00	1.00	0.50	2.00	2.00	1.50	12.00
	Cardamom	1.00	2.00	1.50	1.00	2.00	2.00	1.00	1.00	1.00	1.00	2.00	15.50
	Ginger	1.00	1.00	1.00	1.00	0.00	1.00	1.00	1.00	2.00	2.00	2.00	14.00

Fig-1: Flow chart of wine preparation from Pomegranate:

Pomegranate Fruits

Washing

Aril Separation

Squeezing of arils

Juice Extraction

Addition of Potassium metabisulphite

Keep for 4 hours and raise the Brix to 24

Fermentation at 26^{0}C + 2 for 9 days

Addition of wine yeast culture (50%), Separation of wine by cheese cloth

Filtration

Scaling of filled bottles

Pasteurization at 65 degree Celsius for 20 minutes

Storage of wine

The wine scoring less than 10 marks is considered as poor, between 11 to 15 as better quality and above 15 are treated to be the best quality wines. The dry and sweet wines fetched 15 and 16 points indicate the accuracy of cost effective technology developed for the production of better quality wine. The Vermouth with Cardamom was found to be more acceptable than with clove & ginger.

Wine and health:

1 An occasional wine drinking or a glass, of wine a day is said to stay off near degenerative diseases such as Alzhemers and Parkinsons.

2 Alcohol concentration in wine may not be anti-microbial but phenolic compounds present in wine are anti-microbial and helps in preventing many diseases like dysentery, diarrhea and have protective effects against cardiovascular diseases.

3 There is positive correlation found between moderate wine consumption and improved health possible due to decrease in the prevalence of age-related muscular degeneration.

4 The consumption of moderate amount of wine can reduce the risk of certain cancers and increased consumption may provoke some cancers.

5 Moderate wine consumers may live on an average 25 years longer than teetotalers and considerably longer than heavy drinkers.

Tips to Wine Drinkers:

1 Most home made wines are good to have, at least a year old. They normally stay well for 3-4 years.

2 Like beer bottle it is not necessary to finish your bottle of wine the day you open it. It keeps well for 2-3 weeks with air tight stoppers.

3 Never keep your wine bottle at room temperature. The ideal temperature is between 12 to 15°C.

4 The bottle can simply be kept in the refrigerator.

5 Occasional drinking or a glass of wine a day is found to be good for health.

Fermentation:

The pomegranate fruit must is filled in Laboratory fermentor/flask to its 3/4th capacity and inoculated with 5% wine yeast culture inoculum. The fermentor is plugged with cotton plug to allow sterile air for the initial growth of wine yeast cells. The cotton plug is replaced after 24 hours with air-tight rubber cork fitted with rubber tube. The other end of the rubber tube is kept in water to allow fermentation in the fermentor. The fermentation is better carried out at 21-24°C to get better quality wine.

Siphoning/racking:

After allowing fermentation for 7-8 days, the white coloured and shriveled arils floating in the fermentor is racked with the help of clean muslin cloth to remove pseudo seeds and organic precipitate etc. The fermentation is allowed to continue for another two weeks. The clean wine formed is further siphoned into bottles after removing sediments. The bottles can be stored at 15-16°C temperature or simply be kept in for refrigeration.

Maturation:

The newly-made wine is harsh in taste and yeasty in flavour. The process of maturation or aging makes the wine tasty and fruity in flavour. After storing in refrigerator for one week racking is done and stored back in the refrigerator. Racking is repeated for 2-3 times at intervals of 15-20 days. Racking can be continued further at monthly intervals. During inter-racking times, no head space should be left in the bottles and should be closed tightly. The matured wine after 6 months to a year can be consumed or pasteurized and stored for long time at 15-16°C. The home made wine can stay for 3-4 years.

The detailed production cost of homemade pomegranate wine was worked out and given in Table-1. Vermouth was prepared by addition of different spices like clove, cardamom and ginger at the rate of 0.25, 0.5 and 1.0 g/lit of base sweet wine prepared from fruits of pomegranate. The results of sensory evaluation are presented in Table 3. The results indicated that Vermouth from Pomegranate is found to be acceptable by wine lovers.

The wine prepared from the standard procedure was subjected to organoleptic evaluation for judging its quality based on sensory properties. From the pomegranate fruits (Arakta cultivar) two types of wines were prepared as dry wine and sweet wine with the inoculation of standard wine yeast of *Saccharomyces ellipsoideus* No. 101. The scores as per 20 points developed for organoleptic evaluation are given in Table-2 (Amcrine and Cruncss, 1972)

Scope for Indian Wines:

India process over 6.5 million tons of fruits every year and nearly 30-40 per cent fruits may loose their marketability due to undersized fruits, spotted fruits, cracked fruits, immature fruits and post-harvest losses which accounts to be around 2.5 million tonne. Such losses can be avoided by preparing wine out of such fruits. By considering 50 per cent juice from such fruits, it can be converted to wine as value added product for local markets. A good quality wines can be prepared from Pomegranate, Apple, Pineapple and other non conventional fruits.

Methodology of Wine Making:

Preparation of starter culture:

A standard wine yeast culture of *Saccharomyces ellipsoideus* No. 101 maintained on yeast Extract Peptone Dextrose Agar (YEPT) was transferred into a test-tube containing 5 ml of YEPD broth and incubated overnight at 10°C. The 5 ml yeast culture is further added to 100 ml sterilized grape juice (5% inoculum) in 250 ml flask and kept for incubation at 10°C for 22 hours on a shaking incubator at 120 rpm. This can be used to inoculate the fermentor after activating the yeast cells at 10°C for 1.5 hours.

Preparation of Juice/must:

The fruit of pomegranate were crushed; fleshy and juicy arils are separated by hand. The arils are squeezed to get juice, for further wine making.

Amelioration:

The juice (must) should be tested for its sugar content and pH. If the pH of the juice is found less that 3.2 to 3.3 it can be raised by the addition of Sodium bicarbonate. The TSS of pomegranate juice falls in the range of 14-15^0. Brix could be raised to 22-30^0 Brix by the addition of sugar.

Addition of SO_2:

The addition of potassium metabisulphite (KMS) at the rate of 2.00 ppm (100 ppm SO_2) is added to must, so as to eliminate wine yeast. The 5% cultivated wine yeast is added to the must after having 3-4 hours after the addition of KMS. The colour of the juice is temporarily lost, with the addition of KMS it will slowly regain its colour during fermentation period.

(Courtesy Dr. A. B. Patil, Bijapur)

PACKAGING AND LABELLING

Packing

Fresh pomegranates shall be packed in such a way that the produce is suitably protected during transportation and handling. The containers shall meet the quality, hygiene, ventilation and resistance characteristics to ensure suitable handling, shipping and preserving of the pomegranates. Packages or lot must be free of all foreign material and smell.

The container shall comprise of cardboard cartons of the inter-locking type preferably having a water- proof coating to prevent damage due to high humidity in cold stores.

The mechanical strength of the package should be sufficient to survive pre-cooling/ cold storage procedures and ability to withstand stacking without getting distorted or collapsing during the normal storage life under the storage conditions specified.

The cartons shall be well ventilated and provided with round holes for adequate ventilation.

Specifications details for corrugated fiber board boxes for packaging pomegranate for export as suggested. Boxes conforming to these requirements should be used and which have been treated with wax. Fruits should be packed in cardboard boxes which should be slotted.

The material used for the inside should be new, clean and of a quality such as to avoid causing any external or internal damage to the produce. The use of materials and particularly of papers or stamps bearing trade specifications is allowed provided that printing or labeling has been done with a non-toxic ink or glue.

While packing fresh pomegranates for exports, the packing requirement of the importing countries should also be taken into account.

Palletisation- Pallets of dimensions 1000 mm x 1200 mm and 800 mm x 1200 mm (only for exports to Europe) should be used. Pallets should be strong enough to ensure transit till the final destination. These should be secured by plastic or fiber corner boards. Strapping and clips used should preferably be of plastic material.

Presentation – In addition to the requirements given above, the criteria for presentation as given below shall also be followed in case of exports to the European Union.

a. The contents of each package must be uniform; each package must contain pomegranates of the same origin, variety, class and degree of ripeness.

b. The visible part of the contents of each package must be representative of the entire contents.

Labeling

Each package shall bear the following particulars in letters grouped on the same side, legibly and indelibly marked and visible from the outside.

a. Identification

Packer	Name and address or official issued or
And /or	accepted code mark
Dispatcher	

b. Nature of produce

Pomegranates if the contents are not visible from the outside.

Name of the variety

c. Origin of produce

Country of origin and optionally district where grown, national, regional or local place name.

d. Commercial specification

Class.

e. Official control mark (optional)

f. Net weight

g. Date code

All information shall be on the same side of the package. The marking may be done by using an ink stamp, by printing on to a package, by means of a label firmly fixed to the package or by a combination of these methods.

It is recommended that the information is marked on both ends of the package if it is rectangular or on the two opposite sides if it is square.

The recommended storage temperature may also preferably be printed on the package.

In addition to the requirements given above, the labeling requirements of the importer shall also be taken into account.

SPECIFICATIONS FOR CORRUGATED FIBRE BOARD BOXES FOR PACKING POMEGRANATES FOR EXPORTS

Usually for packing pomegranates for export purpose, a cardboard corrugated fiberboard box of 4.0 or 5.0 kg capacity is used. The dimensions of such boxes depending upon the capacity are:

i. 375 x 275 x 100 mm

ii. 480 x 300 x 100 mm

The detailed specifications are given below:

Sl. No.	Specifications	Ring and Flap Tuck in type	RSC Regular Slotted Container	Slide Type
1	Material for construction	5 ply CFB	5 ply CFB	5 ply CFB
2	Gram mage Gm/m² outer to inner	* 230 x 140 x 140 x 140 x 140	* 230 x 140 x 140 x 140 x 140	* 230 x 140 x 140 x 140 x 140
3	Bursting strength Kg/cm²	Min 10.00	Min 10.00	Min 10.00
4	Puncture resistance Inches/teat inch	Min 250	Min 250	Min 250
5	Compression strength kgf"	Min 350	Min 350	Min 350
6	Cobb value (30 ming/m²	Max 130	Max 130	Max 130

* Outer ply of white duplex board

MAINTENANCE OF RECORDS

Record shall be clearly written, dated and signed by the person entering the data. In case of any change in the record, these shall be signed by the authorized person. All records shall be filled carefully and be retrievable and available for inspection at any time by any person authorized to have access to the records.

Time period for storage of various records should be clearly laid down. The length of storage period depends on the importance of the record relative to the product. No record should be destroyed within the period of a product being marketed and consumed. Pesticide application records and pesticide residue data records needs to be stored for longer periods. Records of equipment inspection, calibration and internal audits also required to be maintained for longer periods.

A recommended list of records essential to be maintained by the pack-house and the time period for holding the record is given at as follows:

LIST OF ESSENTIAL RECORDS

Sl. No.	Record	Time period
1	Management review records	2 years
2	Customer order records	1 years
3	Records of contract review	1 year
4	Records of approved producers and other suppliers	1 year
5	Records of raw material-conditions of supply agreement	3 year
6	Final field survey record	6 months
7	Raw material inspection records	6 months
8	Raw material quality control report	6 months
9	Equipment inspection records	3 years
10	Production records	6 months
11	Final product inspection records	6 months
12	Calibration records	3 years
13	Pesticide application data	1 year
14	Pesticide residue records	3 years
15	Training records	As per policy
16	Internal quality audit records	3 years

SECTION 6. TRAINING

A systematic and well structured procedure should be developed for identifying training needs and providing for training of farmers, exporters and pack-house personnel.

FOR FARMERS AND FIELD STAFF OF EXPORTERS OF POMEGRANATES

a. For producing export worthy pomegranate of good edible quality, free of diseases, insect injury, free of blemishes and also for proper maintenance of an orchard from hygiene point of view.

b. Use, storage and application of pesticides.

c. Techniques of pesticide residue analysis.

FOR FIELD STAFF AND PACK-HOUSE PERSONNEL

a. Raw material assessment.

b. Final and In-process assessment.

c. Processing operation including washing, waxing, fungicide application, drying, preparation of disinfectant, fungicide and wax solution, etc., safe handling techniques of chemicals and probability of contamination and microbial growth.

d. Sizing and grading.

e. Packing methodology.

f. Storage conditions of fresh pomegranates.

g. Hygienic requirements- techniques of hygienic handling and storage of fresh pomegranates.

h. Awareness and responsibilities-Personnel associated with packing should be aware of good management practices, good hygienic practices and their role and responsibility in protecting fresh pomegranates from contamination and deterioration. Packers should have necessary knowledge and skills that minimizes the potential for microbial, chemical and physical contamination.

i. Use of equipment and methodology of quality analysis .

j. Equipment calibration.

h. Training records shall be maintained at both individual and organizational level.

DOCUMENT AND DATA CONTROL

Documents and data including standards, procedures, work instructions, proforma, etc. shall be under control. The control should include the following:

a. All documents requiring control shall be identified and a master list of these shall be prepared and be readily available.

b. The location for each document shall be identified and the current relevant issues of these should be available at identified points.

c. Obsolete documents should be removed from all points of use.

d. Changes made in documents should be approved by the nominated authorized person.

Confidentiality – Many files such as those containing customers specifications are of confidential nature. The confidentiality of these shall be maintained. The supervisors or workers may be issued an outline specification and the original kept securely.

A list of documents which need to be under control is given below:

- Company manual and procedures
- Customers specifications
- Legislation of importing country
    - Pesticide residues, heavy metal limits toxic residues, fertilizers applications, etc.,
    - labeling requirements
    - Weight/volume requirements
- Cleaning schedule for pack-house
- In-process quality control worksheet
- Calibration methods for various equipment

SECTION AND CODE OF HYGIENIC PRACTICES FOR POST-HARVEST HANDLING OF FRESH POMEGRANATES FOR EXPORT

Due to year-round availability of fresh fruits from a global market, there is a substantial increase in consumption of fresh fruits and vegetables in the past two decades. Recent increase in report of food borne illness associated with fresh fruits has raised concerns from public health agencies and consumers about the safety of such farm products.

It is with this back ground, we are dealing with Code of Hygienic Practices both at pre-harvest i.e., primary production and at post-harvest handling stage. The Code of Hygienic Practices concerning primary production of fresh pomegranates has been covered in the pre-harvest manual for production of fresh pomegranates for export.

Code of Hygienic Practices at post-harvest processing, handling and packing stage are discussed in this chapter.

Packing establishment; DESIGN AND FACILITIES

Hygienic aspects in pack house design etc. have been covered in Section 2 of this Manual i.e., Process control and management of quality during production: All aspects concerning hygiene have been discussed under sub section 2.2 pack house requirements.

Control of operation

To avoid contamination from microbial and chemical hazards, entire operations i.e., both at pre-harvest as well as at post-harvest level have to be monitored. Every possible food hazard which can cause contamination of fresh pomegranates has been considered at pre-harvest stage and possible preventive and control systems have been described in chapter 2 of pre-harvest manual for production of fresh pomegranates for export. At post-harvest stage possible hazards which can cause contamination are described in the next paragraphs.

Key aspects of hygiene control systems

Temperature control system

- The details of temperature control system for maintaining hygiene and producing wholesome fresh pomegranates have been described in section 2 process control and management of quality during production.

Post-harvest water use

- Packers should follow good management practices to prevent or minimize the potential for introduction or spread of pathogens in the processing water. As far as possible only potable quality of water should be used. Guidelines for hygienic reuse of processing water are as follow:
- Post-harvest systems that use water should be designed in a way that product does not lodge and dirt should not be allowed to be built up.
- Disinfectants should only be used where absolutely necessary. The disinfectant levels should be monitored and controlled to ensure that effective concentration is maintained. After disinfection, washing with clean water is absolutely essential to ensure that chemical residues don't exceed levels as recommended by Codex Alimentations Commission.
- Where necessary, the temperature of post-harvest water should be controlled and monitored.
- Recycled water should be treated and maintained in a way that it does not constitute a risk to the safety of fresh pomegranates. The treatment process should be effectively monitored and controlled.

Chemicals treatments

- Packers should only use chemicals for post-harvest treatments (waxes and fungicides) in accordance with the general standards on Food Additives or with the Codex Pesticide Guidelines. These treatments should be carried out in accordance with the recommendations for the purpose.
- Sprayers or traversing automatic nozzles for post-harvest treatments should be calibrated properly to control the accuracy of rate of application. These should be thoroughly washed when used with different chemicals and different fruits to avoid contamination.

Cooling of fresh pomegranates

- Forced air cooling in case of fresh pomegranates is the use of rapid movement of refrigerated air over fresh pomegranates in cold rooms. The air cooling system should be appropriately designed and maintained to avoid contamination of fresh produce.

Cold storage

- When appropriate, fresh pomegranates should be maintained at low temperatures at recommended levels after pre-cooling to minimize microbial growth. The temperature of the cold storage should be controlled and monitored.
- Condensate and defrost water from the cooling system in cold storage area should not drip on to fresh pomegranates.
- Inside of the cooling system should be maintained in a clean and sanitary condition.

Packing

- Hygienic aspects during packing of fresh pomegranates have been fully covered under section 5 of this manual.

Packing establishment-maintenance and sanitation

Maintenance and sanitation of equipment and pack-house hygiene has been fully covered in sub section 2.3.1 Equipment and pack-house hygiene under section 2 process control and management of quality during production of this manual.

Cleaning schedule of pack-house for pomegranate has also been provided.

Personnel hygiene

All aspects of personnel hygiene in the pack-house have been covered in section 2 as mentioned above, in sub section 2.3.2. and 2.4 of this manual.

Transportation

All hygienic aspects during transportation of fresh pomegranates in conveyances or containers have been covered in sub section 2.5 transportation requirements under section 2 of this manual.

Training

All aspects of training covering awareness and responsibilities, hygienic requirements and training needs of pack-house personnel have been covered in section 6 of this manual.

Documentation and records

Maintenance of Records, Document and Data Control have been adequately covered in sections 56 and 7 respectively. However, supplementary information is enumerated below:

- Where appropriate records of processing, production etc., should be kept long enough to facilitate trace back and food borne illness investigators if required. The period of holding back records could be longer than shelf life of fresh pomegranates. Documentation can enhance the credibility and effectiveness of food safety control system.

- Growers should keep all relevant information on agricultural activities like site of production, information on agricultural inputs given by supplier, lot numbers of agricultural inputs, use of pesticides, fungicides, water quality data, cleaning schedules of equipment, containers etc.
- Packers should keep all information concerning each lot such as information on incoming materials (information from growers), data on quality of processing water, fungicide and was application programmes, cooling and storage temperatures and cleaning schedules for pack-house premises, packing line, equipment, containers etc.

Recall procedures and trace back

When appropriate growers and packers should have trace back programmes to ensure effective lot identification. The system should be able to trace the sites and agricultural inputs involved in primary production and the origin of incoming material at the pack house in case of suspected contamination.

Grower's information should be linked with packer's information so that the system can trace products from distributor to the field. Information that should be included are date of harvest, farm identification and where possible, the persons who handled the fresh fruits (pomegranates) from the primary production site to the packing establishment. (Courtesy : APEDA, New Delhi.)

Kevin and Erik Wilkins (2009) reported the commercial production of pomegranate in California. The planting system range from row 3.6 m x 5.8m and in row spacing 1.5 to 5.2 m with row spacing often determined by equipment preference for other crops. Training systems are typically free from multiple trunks, although single trunk and trellised production is being used. Irrigation system, crop spacing, are often dictated by companion crops found on individual farms, flood irrigation is found on older trees with newer trees planted with micro irrigation systems. Organic productions of pomegranates are limited due to market acceptance; challenges for organic production are *Alternaria alternate* and *Aspergillus niger*, as well as omnivorous leaf roller (*platynota stultana*) and flat mite (*Brevipalpus lewisi*). Numerous herbicides are registered for use in conventional pomegranates and weed barriers and hand weeding organic system. Both fresh market and juice fruit continue to be commercially harvested by hard but attempt at mechanization for both uses are underway. Commonly pomegranate yields are low as 6-10 MT/ha when used exclusively for fresh market consumption, but can be as great as 25-35 MT/ha when grown for juice, with the fresh juice ratio dependent on both market conditions and autumn weather patterns. Pomegranates are also sold through food retailers in North America, and their current preference for high sugar/high acid, dark red rind colour and dark red aril colour favour cultivars such as Granada clonal and Foothill in early season. Variation for fruit characteristics may influence consumer acceptance to other fruit types already common throughout pomegranate consuming cultures.

SECTION-9 INTERNAL QUALITY AUDITS

An internal audit programme shall be established by the pack-house to cover the entire system, facilities and operations. The nature of the programme will depend on the size of operations.

A programme and timetable of audits covering all systems shall be drawn up. Audits shall be carried out by management staff not directly involved with that activity.

A suggested programme of internal audits is as give below:

Sl. No.	Area of Audit	Frequency
1	Pack-house audit	Once every 3 months (a sample audit sheet)
2	Quality Assurance Audit	Once in 6 months (minimum)
3	Audit of entire operations covering the different areas of operation from the production till the final package in and dispatch of the product.	Once a year (minimum)
4	Maintenance of files and records (to cover maintenance of suppliers files, specification, rejection, visit record, etc.).	Once a year (minimum)

Records of the audit shall be maintained for a minimum period of 3 years.

SAMPLE AUDIT SHEET FOR PACK-HOUSE AUDIT

Location____________________ Audit team 1) ____________________

2) ____________________

Date of previous audit ____________ 3) ____________________

Sl. No.	Particulars	Action	
		By whom	By when
1	**Fabric of building** Repairs required (if any) Improvements		
2	**Vermin and bird control** Entry points-doors, windows, etc. Routine controls - Baits in place and numbered - Visit report in file - Baiting plan in file		
3	**Contamination/hygiene risks Toilets** - Cleaning schedule in force - Hand-wash, soap and hand dry facilities in force **Pack – house** - Cleaning schedule in force - Clearance of rubbish - Storage of packaging and chemicals - Water quality - Personal hygiene in force - Work instructions in place - Pack-house disciplines in force **Glass** - Lights protected, etc.		
4	Equipments - Maintenance and condition - Repairs, improvements - Weight checks records - Calibration records		

Report agreed and signed by : 1) ____________________

2) ____________________

3) ____________________

Date ________________

Copies to relevant managers for action ____________________________

SUMMARY OF RECOMMENDATION AND GUIDELINES TO BE FOLLOWED FOR POST HARVEST HANDLING OF POMEGRANATES TO ENSURE QUALITY

Usually, in India post-harvest handlings not taken very seriously. This aspect is very technical and, therefore following methodology needs to be adopted in post-harvest handling of pomegranates:

Selection of pomegranate orchard

It is always best to select an orchard where recommended fungicide, pesticide and balanced nutrients have been applied at the required and appropriate time. Following guidelines should be considered while selecting a pomegranate orchard for export purposes:

- An orchard where last pesticide application has been made at least one month before harvest.
- The orchard should be apparently healthy, free of nutrient deficiencies, free of sunscald, orchard clean and not strewn with fungal or bacterial infected leaves, fruit borer infested fruits and also free of rusting(mite damage) and rough surface on fruits (thrips damage).
- Fruits should be of adequate size and diameter characteristic of the variety.
- Ensure that the fruits have reached the required maturity and colour. The brilliant fruit shine and appearance is most important in terms of value realization foreign countries. This must be looked into while selecting the orchard.
- Some tests are necessary at the field level to find, out quality parameters of fruits while selecting the pomegranate orchards. The tests to be done and the requirement are given below:

i)	Weight	-	Weighing scales
ii)	Size of fruits	-	Plastic rings of different sizes
iii)	T.S.S.	-	Refractometer
iv)	Temperature of fruit	-	Temperature probe

Maturity and maturity standards

It is generally considered that the fruit is ready for harvest when fruits have attained the distinctive fruit colour characteristic of the variety and make a metallic sound on tapping.

Maturity standards vary with the cultivar. For example, Ganesh cultivar is considered mature when T.S.S. reaches 14.5^0 B or so and acidity % comes to a level of 0.3 % or 0.35 % . The maturity indices vary slightly for Mridula cultivar as in this if T.S.S. has reached 16^0 B or so and acidity around 0.45 % it is considered fit for harvest.

Quality requirements of importing countries

In foreign countries especially in European countries where it's demand is maximum, like brilliant shiny crimson or red colour of fruits and sparkling red colour of arils (kernels). This is because whole fruits are widely used for decoration and kernels are used for garnishing desserts and salads. Also juice from dark red kernels is used for syrup making which is further used for flavouring mixed drinks, topping of ice cream and desserts. Therefore, desirable fruit characters of importing countries are enumerated below:

i) Smooth cutting at the stem end

ii) Shining dark rose pink colour of the fruit

iii) Fruit weight around 200-400 g.

iv) Round shape of fruit

v) Uniform size and shape of fruit in a pack.

vi) Dark rose pink arils (kernels)

vii) Softness of seeds

viii) High total soluble solids near about 16^0 B to 17^0 B

ix) Freedom from scars, blemishes, scratches, rusting, rough surface, disease spots and fruit borer

x) Bracts of calyx having freshness and without any damage

xi) Pleasant flavour and aroma

xii) Without blackening of arils

Method and time of harvesting

- Harvest pomegranates either early in the morning or in the evening hours
- Use clippers and bags for harvesting
- Do not pull the fruits
- Cut the stem end close to the base
- Do not bruise the base while cutting the fruits. Handle carefully at harvest
- Do not keep longer stem end, as it can cause damage to other fruits
- Keep the fruits in plastic crates for emptying the bags
- Do not keep fruits on the ground
- Do not keep empty crates in the sun but in shade
- Do not overfill the crates with fruits and keep them in shade and not in open sun

- Do not damage calyx while harvesting or while keeping in crates or during field storage Crates should be washed with water and detergent. Clippers should also be cleaned frequently with a clean wet cloth and dipped in 1 % sodium hypochlorite solution.

Transporting fruit from field to pack-house

- Transport the fruits immediately to pack-house without any loss of time
- Transport should be available on demand to avoid any time lag between harvest and transport to pack-house
- Shiny appearance and bright colour of rind are prime factors for higher value realization and therefore shine and appearance of fruits should not be spoiled during loading, unloading and during transport.
- There should be no jerks or bumps in the vehicle to cause rubbing among the fruits
- The vehicle should be clean to avoid infection to the fruits
- Preferably no other product should be transported along with pomegranates

Receipt of pomegranates at pack-house

- On arrival, the unloading should be done gently to avoid bruising and preferably in the sorting bay having cool temperature to avoid multiple handling of the fruit.
- Do not stack too many crates over each other to avoid tumbling and bruising of the fruits
- Note down both tare and net weights of crates.
- Care needs to be taken that grower's code is maintained and produce from different growers does not mix.

Sorting, washing and treatment with fungicide and wax on a packing line

i) Sorting

Pomegranate fruits in plastic crates are directly put on the inspection roller cum-conveyer. At this point, all fruits showing greenish colour, fruits with sunspots, affected by fruit borer or having disease spots, rusting and rough surface which cannot be accepted in the international market must be eliminated. Also fruits having cuts, blemishes, damages, etc. need to be sorted out manually and rejected.

ii) Washing

After rejecting all unwanted fruits, these must be washed. It is preferable to add a disinfectant like sodium hypochlorite @ 100-200 ppm. After chlorine disinfection, rinsing with clean water is a must. Only potable water should be used for all post-harvest processing purpose.

iii) Elimination of excess water on the fruit surface

After washing, excess water on the fruit surface must be removed. Efficiency of wax and fungicide application depends upon how effectively excess water adhering to fruit skin has been eliminated. This is usually done by making use of absorbent rollers.

iv) Waxing and fungicide application

Most common and effective fungicides are Thiabendazole or Benomyl at 1000 ppm concentration. The concentration of the fungicide will vary with the speed with which fruit passes on the conveyer and the quantity of fungicide deposited on the fruit surface by traversing nozzle. Each packing line has its own speed depending upon design, capacity and manufacturer and thus concentration of fungicide must be calibrated/worked out so that international residue limits are not crossed and there is sufficient deposit to protect from post-harvest losses.

Correct strength of the fungicide is absolutely important to avoid post-harvest losses. It is beneficial to prepare fresh fungicide solutions everyday just before application for maximum effectiveness.

a) Poor quality wax should not be used. Only good quality edible wax need to be used.

b) Monitor edible wax quantity required on packing line for very good shine on fruit surface.

c) Monitor concentration of the fungicide which will have minimum residue on the fruit and minimize post-harvest losses.

d) Effective wax coating is very important especially in pomegranates in shriveling of the fruit. In such a situation, realization value or market acceptability is reduced.

e) While using readymade wax and fungicide solutions, their manufacturing date must be considered.

Drying the wax coating

The wax coating on the fruits is usually dried in suitable drying tunnel of the packing line at a temperature of about 30-35^0 C. However, it must be ensured that:

- Bulbs, heaters, hot air circulation system along with fans, air knives, etc. must work properly to attain proper drying.
- The temperature inside the tunnel must not go beyond 30-35^{0}C

Sizing the fruits

After drying, the fruits are fed into feeder channels of a grader or size which works on the principle of roller sizing which is quite common in grading of fruits and gives good results. In this method, the distance between the rollers can be adjusted for grading different sized fruits.

For pomegranate following grades are adopted:

Sl.No.	Different size ranges (g)	No. of fruits in	
		4.0 kg box	5.0 kg box
i	45 to 500	Not exported	Not exported
ii	400 to 450	-do-	-Do-
iii	350 to 400	10-12	12-15
iv	300 to 350	12-15	15-18
v	250 to 300	12-15	15-18
vi	200 to 250	15-18	18-20

For export purposes mostly 200 to 400 gm fruit weight are used. Weight of 400 to 450 gm and 450 to 500 gm are called 'A' grade fruits which are marketed within India in domestic market. Following steps must be taken in sizing of fruits:

i) Variation in a size grade should be minimum as far as possible
ii) Quality supervisors must ensure that bad quality fruits must not go into the fruits which are to be packed in cardboard boxes.

Nose cleaning

Before packing a box, fruits at stylar end or calyx side are wiped with a cloth soaked in a fungicide solution (Benomyl or Thiabendazole 1000 ppm) to avoid cross contamination in a packed box.

Packing and coding

a. Packing methodology

For packing mostly 4.0 to 5.0 kg box is used and fruits 10to 20 in number are packed in a box depending upon the fruit size. Usually a 5.0 kg box contains 10-15 fruits of Ganesh cultivar, whereas fruits of Mridula,Arakta & Bhagwa 18-20 in number are accommodated. The dimensions of a 4.0 kg box are 375 x 275 x 100 mm and that of 5.0 kg box are 480-300 x100 mm. For Gulf countries, paper shreds below the fruits, in between and on the top of the fruits are used in packing to counteract any bruising during transit However, for European market, cardboard partitions are used in packing to avoid movement of fruits within the pack and thereby bruising. To further safeguard the fruits from injury, bubble plastic is used and placed at the bottom, sides and at the top of the fruits.

The size of the box can be designed according to the requirements and accommodating a particular fruit size which is liked in the importing country. However following suggestions must be taken into consideration.

i) Cardboard boxes must have ventilation holes, hand holes, etc.

ii) The packing should be such that there should not be pressing of any fruit so as to cause bruise on the skin.

iii) Placement of fruits by putting accurate sized fruits should be such that there is no movement of fruits inside the pack.

iv) The pallets should be of dimensions namely 1000 mm x 1200 mm. For exports to Europe, pallets of dimension 800 mm x 1200 mm should be used.

v) Only plastic material should be used for strapping and clipping for securing the cartons to the pallets.

Boxes should be coded for date of packing, product's name, grower's code for tracing its origin. The box must be labeled with a very attractive label and should contain following information:

- Name of the produce
- Variety
- Origin of produce
- Commercial specification of class
- Date of packing
- Name of packing organization
- Gross weight/net weight
- Number of fruits/fruit size
- Recommended storage temperature
- Additional labeling requirement of importer should also be included

Pre-cooling

Pomegranates should be cooled to 5-7° C temperature as soon as possible and at least within 10 hours after harvesting.

Before storing at above temperatures, fruit needs to be pre-cooled by forced air. Pre-cooling is very important for successful sea shipping. The ideal pre-cooling temperatures are the same, i.e., 5-7°C. There can be chilling damage if fruit is placed below 5° C for long time. The humidity should be around 90-95 % during pre-cooling. Following precautions may be taken during pre-cooling:

- Prompt pre-cooling is very important for avoiding post-harvest losses and distant transportation.

- The filled cartons should be shifted to pre-cooling chamber without any delay.
- On transfer of the product, the outflow temperature and humidity should be properly recorded.
- The products should remain in the pre-cooling chamber till the temperature of the fruit reaches to the required stage. Do not allow to fall the temperature otherwise chilling injury will make the fruits unfit for consumption.
- The temperature, humidity and air circulation of the room shall be properly maintained and checked at pre-determined intervals.
- After pre-cooling and Palletisation, strapping should be done. This needs to be done only in cold room to maintain the cool chain.

Cold storage

The ideal storage temperatures are 5 – 7 ^{0}C. The above temperatures need to be kept throughout the transport (for destination to importing country) and further storage there. The temperature should never go below the figures stated, otherwise it is likely to result in chilling injury.

The relative humidity should be kept at 90-95 % throughout transport and storage.

Susceptibility to chilling also varies with maturity and variety. Pomegranates subjected to too low temperature, develop brown discoloration of skin, pale colour of arils and brown discoloration of white segments separating the arils.

It is critical for pomegranate storage that cold storage should be uninterrupted. If there is interruption, it negates all prior careful handling of fresh product.

Quality appraisal

Quality checking should be done with regard to following parameters:

- Cutting stem end to appropriate size
- Fruit size/uniformity
- Colour of the fruit/uniformity
- Condition of calyx
- Freedom from sunscald, disease spots, fruit borer, rusting and rough surface
- Fungicide residue in no case should exceed prescribed limits
- T.S.S. (^{0} Brix)
- Spoilage at the port due to diseases before dispatch to importing countries.

Level for quality checking

For ensuring export quality fruits, quality checking is suggested at following levels:

- Pack-house before dispatch to the port
- At the port before dispatch to the importing country
- The importing country. This feedback is very important for further improvement. Acceptability and grading given in the importing country is very important for managing future exports.

Equipment required for quality control

- Refractometer – T.S.S.
- Burette, pipette, volumetric flask, glass rod, NaOH, Phenolphthalein for acidity analysis
- Scales for recording weight
- Pressure tester to test the firmness of the fruit
- Temperature probe to record temperature inside the fruit
- Spectrophotometer/HPLC to determine residue levels* of fungicides in the fruit.

Container loading

- In order to maintain the cool chain the temperature of the loading bay should be controlled preferably to 20 – 25 °C.
- The transfer of pomegranates from the cold store to the loading bay and to the container should be smooth and quick so that it does not result in temperature increase of pomegranates.
- Before shifting of pomegranates to containers of Reefer vans, ensure the appropriate recommended temperature and humidity inside the container.
- The doors of the containers must be closed before and after loading pomegranates.
- Ensure that temperature and humidity recorders are placed inside the container.
- The cartons/pallets in the container are to be arranged in such a manner that there is no movement within the container. This can be done by placing the pallets in an interlocking manner or by use of iron rods/ pegs for plugging the empty space.
- It is preferable or rather should be ensured that only a single type of product is loaded to one container.

Shelf life experiments

A few cartons should be kept aside to check the shelf life of the product and should be tested periodically to ensure that the quality is maintained till it reaches the customers/consumers. Separate areas in the pack-house/cold store shall be set aside for the shelf life experiment boxes.

Implementation of code of hygienic practices at post-harvest handling stage of fresh pomegranates.

- Design and facilities of pack-house should be such that there is no ingress of vermin from birds, etc. It should be protected by fence and flooring, etc. to enable proper cleaning.
- Water for washing and especially rinsing should be of potable quality to minimize potential for introduction of pathogens, etc.
- Disinfectants wherever necessary should be monitored and residues should not exceed levels recommended by CODEX.
- If recycled water is used, it must be treated in a way that it does not become a risk to the safety of fresh pomegranates.
- Use only those waxes, fungicides at post- harvest stage, which are safe and approved by CODEX and should be in accordance with the recommendations.
- Sprayers and automatic nozzles should be calibrated to control the accuracy of rate of application and properly cleaned to avoid contamination.
- The air cooling system be designed and maintained in a manner to prevent contamination of fresh pomegranates.
- The temperature of cold storage should be controlled and monitored to minimize microbial growth.
- The cooling system should be maintained in a clean and sanitary condition.
- Cleaning schedule of equipment and pack-house should be strictly implemented.
- All aspects of personnel hygiene in the pack-house should be strictly enforced.
- All hygienic aspects during transportation of fresh pomegranates in Reefer vans or containers must be looked into and no room for contamination needs to be left.

(Courtesy Post Harvest Manual for export pomegranate, APEDA, 2002, New Delhi)

Chapter 12

Insect Pest of Pomegranate

POMEGRANATE BUTTERFLY

Virachola isocrates (Fabricius) is economically the most important of all the butterflies, perhaps the only one that is constantly and regularly injurious. It is a polyphagous pest having a very wide range of host plants, including, aonla, apple, ber, citrus, guava, litchi, loquat, mulberry, peach, pear, plum, pomegranate, sapota and tamarind. It is widely distributed all over India and is found wherever pomegranates are grown.

The female lays eggs singly on calyx of flowers or small fruits. On hatching, the caterpillars bore inside the developing fruits and are usually found feeding on pulp and seeds just below the rind. As many as eight caterpillars may be found in a single fruit. Subsequently, the infested fruits are also attacked by bacteria and fungi causing the fruits to rot. The conspicuous symptoms of damage are offensive smell and excreta of the caterpillars coming out of entry holes, the excreta are found stuck around the holes. Sometimes the holes may also be seen plugged with the anal end of a caterpillar. The affected fruits ultimately fall down and are of no use; even if the fruits be picked before falling down; these fruits have no market value.

Eggs are shiny white in colour and oval in shape. Full grown caterpillars are stout, 17 to 20 mm long, dark brown in colour with short hair and whitish patches all over the body. Just before pupation, the caterpillars come out of the fruits and tie the stalk of fruit with main branch of the tree with fine silken strands to ensure that the fruit does not fall down, then re-enter the fruit and pupate there in. Occasionally the caterpillars may pupate outside also, attaching themselves to the stalk of the fruits. Adult butterflies are medium sized, glossy-bluish-violet (males) to brownish violet (females) in colour; females have conspicuous orange patch on the fore wings. Wing expanse is 40 to 50 mm.

Incubation, larval and pupal periods range from 7 to 10, 18 to 47 and 7 to 34 days respectively with four overlapping generations in a year. The pest breeds throughout the year.

To reduce the incidence of this pest, remove and destroy all the affected fruits. The only other effective though expensive method is bagging of fruits and may be practiced, if the number of fruit trees is limited. Alam (1962) from Bangalore suggested spraying with 0.03 % phosphamidon just when the fruit formation starts.

Pomegranate fruits are also damaged by pomegranate borer, *Deudorix epijarbas* (Moore), castor capsule borer, *Dichocrocis punctiferalis* Guenee, pyralid borer, *Euzophera punicaella* Moore and fruit sucking moths, *Otheris cajeta* (Cramer) and *O. materna* (Linnaeus). Besides, Ayyar (1924) reported two pentatiomid bugs, *Jurtina indica* Dallas and *Halyomorpha picus* (Fabricius) puncturing the tender and ripe fruits. These cause only minor damage to pomegranate fruits. Except *D. epijarbas*, others are polyphagous pests and cause major damage to other economic crops. *D. epijarbas* has been occasionally causing serious damage, specially in Himachal Pradesh and Uttar Pradesh.

BARK EATING CATERPILLARS

Bark eating caterpillar: *Indarbela tetraonis* are widely distributed in the tropical and sub-tropical regions of Indian sub continent. These are highly polyphagous species attacking a wide range of fruits like pomegranate. The nocturnal female lay eggs in groups of 15 to 25 in cracks and crevices or under loose bark of stem, trunk, stem or main branches usually near the forks.Upto 2000 eggs are laid by a single female mainly during summer (May – June). The incubation period is 8 to 11 days and the new hatched dirty brown caterpillar nibbles the live bark superficially for 2 to 3 days before boring in to the branch/stem. The larvae takes as many as 9 to 11 months to complete development. Larval feeding and development slow down during winter months. The full-grown larva is light brown or dirty brown with strong biting mouth parts. It pupate in the bored hole in April and pupal stage lasts 3 to 4 weeks. Adult moth emerge in May – June leading a portion of empty pupal case protruding out of the bore hole. The moth is stout, pale brown with rufous head and thorax. The forewings are pale brown with numerous brown spots and bands/streaks while the hind wings are fucous. Male are slightly smaller than female. Adults have a short life span of less than a week. There is only one generation in a year. The larva while feeding on bark destroy the food transporting phloem vessels and thus interrupt the flow of nutrients and sap. Further boring in to the branches and stems by the larvae also weakens trees. All these in turn affect the vigour, vitality, and fruiting capacity of the trees. Severely infested branches/stems even main trunk often breaks off under the pressure of strong winds. The caterpillars construct the silken gallaries while feeding on bark, such tunnel extend around the feeding sites and bore hole in zigzag manner usually near the fork of branches and may be a few to more than 65 cm in length, and thus can be easily located, tunneling downward and feed on the bark sheltered under the cover of webbing during night. There is only on larva in each hole or webbings may be present on a tree.

Management:

The pest is indeed very difficult to control provided timely and appropriate management practices are adopted. Keep the orchard clean and avoid over crowding of trees, kill the caterpillars mechanically by inserting thin wire or iron spike in to the larval holes. The mechanical destruction of the larva rather tedious on tall trees and in large area is quiet effective. In early stage of infestation i.e., September – October clean the webbed areas and paint the bark around borer holes with emulsion prepared by adding 10 ml Monocrotophos or injecting the dichlorovos 30 ml/liter in the holes and plugging with wet mud. During February – March remove the webbing and give the following treatments inject in to each borer hole 5 ml emulsion prepared by diluting 20 ml dichlorovos or 10 percent kerosene or injecting petrol. Push into each hole a cotton swab soaked in any one of the above-mentioned insecticide.

Mites (*Aceria granti)*

The pomegranate plant has been found affected extensively by *Aceria granti*. This eriophyid mite is yellowish and with robust body. This mite causes an interesting leaf roll, which consists of a downward tight rolling and curling of the whole length of the leaf margin into tube. Eventually the infested margin rolls towards mid rib but does not cover it. The mites are wrapped in the thickened and turned over leaf edge. Infestation is so heavy that every young leaf is rolled, distorted and twisted.

Control

Spray 0 mite 57 EC @ 2 ml per litres of water evenly after thorough mixing.

STEM BORING BEETLES

Celosterna spinator Fabricius and *Olenecamptus bilobus* Fabricius have been reported boring the stems and trunk of pomegranate trees. Both are polyphagous pests causing minor damage. *O. bilobus* is widely distributed in the Oriental region. It is primarily a pest of *Ficus spp.* It prefers breeding in dead wood but also attacks the living branches. The grubs bore inside the trunk and feed on sapwood. Adult beetles are active by day and feed by gnawing the green bark of shoots. The life-cycle is annual with an extended emergence period from May to November.

C. spinator is a major pest of babul, *Acacia arabica*. Adult beetles have pale yellowish-brown body with light grey elytra and are 30 to 35 mm long. These are nocturnal in habit and appear with onset of monsoon; emergence is accelerated by continuous or heavy rains. Egg period is 12 to 15 days, grub 9 to 10 months and pupal period 16 to 18 days. There is only one generation in a year and longevity of beetles is 45 to 60 days.

Generally no control measures are required. Nevertheless, the measures suggested against bark eating caterpillars will prove effective in controlling these beetles as well.

SAP SUCKING INSECTS

Drosicha mangiferae (Green), *Hemiaspidoproctus cinerea* (Green), *Icerya purchasi* Maskell, *Planococcus lilacinus* (Cockerell), *Andaspis hawaiiensis* (Maskell), *Aonidiella orientalis* (Newstead), *Duplaspidiotus tesseratus* (de Charmoy), *Hemiberlesia punicae* (Signoret), *Lindingaspis greeni* (Brain and Kelly), *L. rossi* (Maskell), *Parlatoria oleae* (Colvee), *Pinnaspis theae* (Maskell) are some of the mealy bugs and scale insects reported from various parts of India on pomegranate leaves. Most of these are polyphagous, having a wide range of host plants. Due to damage by these insects, the trees are devitalized, resulting in shedding of buds and flowers and smaller sized fruits. To prevent the attack from spreading, prune and destroy the affected parts in the initial stage of attack. If the infestation becomes severe, spray with 0.04 % diazinon or monocrotophos.

Pomegranate aphid *Aphis punicae* and pomegranate whiteflies *Siphoninus finitimus* Silvestri and *S. Phillyreae* (Haliday) are also minor pests recorded occasionally. They feed by sucking the cell sap from leaves and tender twigs. The affected parts get discolored and disfigured. These insects also secrete copious amount of honeydew on which sooty mould develops, hindering the photosynthetic activity of the plant.

Thrips, *Retithrips syriacus* (Mayet) and *Rhipiphorothrips creuentatus* Hood have been recorded feeding on leaves, while *Anaphothrips oligochaetus* Karyny. *Ramaswamiahiella subnudula* Karny and Scirtothrips dorsalis Hood infest the flowers. These are all minor pests of pomegranate. Nymphs and adults lacerate the leaves, flower stalks, petals and sepals and rasp the sap that oozes out of these wounds. As a result, the leaf tips curl and dry away while flowers are shed and ultimately the fruiting capacity of the tree is adversely affected.

To control the aphid, whiteflies and thrips spray with 0.03% dimethoate, oxydemeton methyl. Two spraying at an interval of 10 to 15 days will be sufficient to keep complete check of these pests.

LEAF EATING CATERPILLARS

Leaves of pomegranate trees are often attacked by one or the other lepidopterous larvae. Those of regular occurrence include castor semilooper. Hairy caterpillars, slug caterpillars and bag worms.

Castor semilooper, *Achaea janata* (Linnaeus) is an important pest of castor. It has also been reported damaging ber, citrus, grapevine, guava, pomegranate, etc. The pest has country wide distribution. Eggs are laid singly on tender leaves usually on ventral surface, one to six eggs may be found on a leaf. A female lays on an average 400 eggs. Freshly hatched caterpillars congregate on the leaves of various weeds and a few economic crops, including pomegranate, ber and castor, and feed on the chlorophyll. Later, they segregate and feed voraciously devouring the entire leaf lamina. In case of severe infestation, the young trees may be completely defoliated. The moths are also destructive. They suck the juice of various fruits, like citrus, grapes, guava, mango, etc. Eggs are hemispherical, bluish-green and ridged with 40 to 45 striate. Caterpillars are semiloopers about 5 mm when freshly hatched becoming 55 to 65 mm when fully grown.

The caterpillars show conspicuous colour variations some are grey with red or brown lateral stripes while others are bluish-grey specked with blue-black and having yellowish lateral stripes. Pupation takes place in the soil 40 to 50 mm deep or even among the fallen leaves or in folds of leaves. Freshly formed pupae are glistering dark green, later becoming brown. Moths are stout, pale reddish-brown with wavy lines on fore wings and black hind wings having a medial white band and three large white spots on outer margin. Anal and apical margins of the wings are fringed with hair. Wing expanse is 50 to 65 mm.

Incubation period is 3 to 5 days while the larval and pupal periods last for 9 to 23 and 7 to 26 days respectively. Preoviposition period of moths extends from 6 to 21 Days and a life-cycle occupies 3 to 5 weeks during active period of the pest. There are five to six overlapping generations in a year and over wintering takes place in pupal stage.

To check the damage caused by these semiloopers, collect and destroy the caterpillars mechanically. In case of severe infestation, dust 5 % Malathion or spray with 0.05 % dichlorvos or endosulfan. This will also kill other leaf eating caterpillars, if any .

Ber hairy caterpillar, *Euproctis flava* (Bremer), plum hairy caterpillar, *E. fraterna* (Moore) and castor hairy caterpillar, *E. lunata* Walker are the species reported damaging pomegranate trees. These are highly polyphagous pests, found all over Indian sub-continent. Eggs are laid in clusters on ventral surface of leaves and covered with hair. On hatching, the caterpillars feed gregariously on epidermis of leaves; later they segregate and feed voraciously, defoliating the entire trees. Pupation takes place in hairy cocoons on leaves or on branches. Moths of *E. flava*, have head and thorax bright orange-yellow and abdomen paler. Forewings are orange yellow in colour with orange spot at end of cell and three sub apical black spots; hind wings are paler than forewings. Wing expanse is 30 mm (0) to 40 mm (0). Moths of *E. fraternal* are more or less similar to those of *E. flava*, except that these are smaller in size, wing expanse being 24 mm (0) to 34 mm (0). *E. lunata* moths are pale in colour with white forewings having a large black lunule; wing expanse is 34 mm (0) to 38 mm (0).

Biology of these species is more or less same. The eggs hatch in 5 to 10 days, larval period lasts for 4 to 5 weeks and pupal stage occupies 10 to 12 days-thus entire life-cycle is completed in 45 to 57 days.

Porthesia scintillans is another hairy caterpillar widely distributed in Indian subcontinent. This is also polyphagous pest, damaging, apple, mango, pomegranate, etc. Caterpillars are dark brown with a series of crimson lateral tubercles on a yellow line bearing tufts of grey hair. Moths are yellowish with reddish line and spots on the edges. Forewings are vinous-brown irrorated with dark scales and hind wings are fuscous-brown with broad yellow margin; wing expanse is 20 to 26 mm and 32 to 38 mm in case of males and females respectively.

Egg, larval and pupal periods last for 6 to 10, 30 to 40 and 8 to 12 days respectively.

Arcyophora dentula Hampson has been reported as a minor pest. The caterpillars gnaw holes in the leaves and ultimately defoliate the tree. Maximum damage is caused during September-October.

Caterpillars of *Creatonotus gangis* (Linnaeus) have been reported feeding on leaves during January to March in Andhra Pradesh.

Egg, larval and pupal periods last for 3 to 4, 30 to 33 and 10 to 13 days respectively.

Dusting with 10 % BHC is effective for checking these hairy caterpillars. It is desirable to carry out the campaign on a cooperative basis and during early stage of infestation when the caterpillars are feeding gregariously; at this stage, even 5 % BHC dust will give a good mortality of these caterpillars.

Bagworms are widely distributed all over India. The species recorded feeding on pomegranate leaves, include, *Clania crameri* Westwood and Acanthopsyche species. The caterpillars construct a case over their body and live within, nibbling leaf lamina. Females are apterous and devoid of antennae, mouth parts and legs. Males are winged and have bipectinate antennae and short mouth parts.

BEETLES AND WEEVILS

Anomala dimidiate Hope, *Hoplasoma sexmaculata* Hope, *Mimastra cyanura* Hope, *Myllocerus laetivirens* Marshall and M. *undecimpustulatus maculosus* .Desbrocher have been occasionally reported feeding on pomegranate foliage. These are polyphagous pests, destructive mostly to temperate fruit trees and of minor importance in case of pomegranate. The grubs live in the soil and feed on the roots and other organic matter. It is only the adults that come out of the soil at night and feed on foliage. Generally no control measures are required against these pests, but if and when necessary, dust with 5 to 10 % BHC dust.

FRUIT FLIES

Mango fruit fly, *Dacus dorsalis* (Hendel) and peach fruit fly, *D. Zonatus* (Saunders) have been reported attacking pomegranate fruits as well, though the loss caused is of minor importance. These flies are cosmopolitan in distribution and feed on a vast variety of fruits and vegetables. Both the species are found attacking apple, bael, ber, citrus, fig, guava, mango, peach, pomegranate, sapota, etc. In addition, D. *dorsalis* has also been recorded on apricot, banana, jack-fruit, loquat, pear, persimmon, plum, etc., while *D. zonatus* has been recorded damaging custard apple.

Eggs are laid just below the epidermis of ripening fruits. The affected fruits show dark brown punctures through which juice oozes out. On hatching, the small white grubs feed inside the fruits, as a result of which the fruits rot and fall down.

To prevent population build-up and carry over of these flies, collect and destroy promptly all the fallen and affected fruits. To avoid the infestation, harvest the fruits before they ripen as the flies are not able to puncture the hard rind of unripe fruits.

Othreis fullonia *(Clerk)*

HOSTS

The Pacific fruit-piercing moth attacks many fruit and vegetable crops. Fruit crops attacked include apples, apricots, bananas, breadfruit, coffee, figs, grapefruit, guava, kiwifruit, litchi, longan, mandarins, mangoes, nectarines, oranges (especially navels), papaya, passion fruit, peaches, persimmons, pineapple, plums, and star fruit. Vegetable crops attacked include tomatoes and melons (Waterhouse and Norris, 1987).

Except in the Pacific, larvae develop on plants belonging to the Menispermaceae family, especially the creepers belonging to the Tinospora, Tiliacora, Triclisia, and the Stephania genuses (Banziger, 1982; Comstock, 1963; Hargreaves, 1936; Mosse-Robinson, 1968). In the Pacific, the larvae of this moth almost exclusively develop on plants belonging to the Erythrina genus, or false wiliwilis, a common landscaping plant, the only exception being the creeper, Stephania forsteri (Menispermaceae) (Cochereau, 1977). In Hawaii larvae have been found on Erythrina fusca and Erythrina variegata orientalis (Heu *et al.*, 1985).

DISTRIBUTION

This moth is native to the Indo-Malaysian region, and widespread throughout the Pacific basin, Asia, and Africa. It is not recorded in the Americas. Some countries where it is found include Australia, China, India, Japan, Korea, the Philippines, Papua New Guinea and Thailand. Refer to Waterhouse and Norris (1987) for an extensive list of countries where this moth is present. It was first reported in Hawaii on Oahu in 1985, and by 1986, it was present on Kauai, Hawaii, Maui and Molokai.

DAMAGE

For most moth and butterfly pests, the caterpillars are the damaging stage. The Pacific fruit-piercing moth differs in this aspect because it is the adult moth that is the damaging stage, and the larvae are essentially not harmful. The mouth parts of the moth are about an inch (2.5 cm) long and strong enough to penetrate through tough-skinned fruit. Once the moth has punctured the skin of the fruit, a process that usually takes a few seconds, it feeds upon the juices of the fruit. Feeding occurs at night and the fruit does not have to be ripe to be fed upon by this moth (Waterhouse and Norris, 1987). Fruit flesh damaged by this moth becomes soft and mushy differing from fruit damaged by fruit flies which is more liquid (Heu *et al.*, 1985).

Damage caused by this pest is not only a result of the direct feeding of this moth but also by the fungal and bacterial infections that develop at the wound site. This moth is a known vector of Oospora citri, a fungus that rots the fruit and has a penetrating odor that attracts this moth. Other microorganisms that gain entrance into the fruit and cause rotting include *Fusarium* sp., *Colletotrichum* sp. (Banziger, 1982), and several types of bacteria (Hargreaves, 1936). When moths are abundant green fruit is attacked, causing premature ripening and dropping of fruits (Kumar and Lal, 1983). On oranges, a green fruit turns yellow at the site of the piercing and fungi soon develop within the wound (Waterhouse and Norris, 1987). On tomatoes, the puncture of the tomato skin causes the fruit to turn white and quickly rot (Comstock, 1963).

Incidence of damage by this moth is normally low, however when outbreaks occur, most of the crop is affected.

BIOLOGY

The duration of the life cycle from egg to egg-laying adult female was 35 to 49 days in a study conducted in Fiji at temperatures between 82.4° to 91.4° F (28° to 33° C) (Kumar and Lal, 1983) and 30 to 33 days in the warmer conditions of new Caledonia (Waterhouse and Norris, 1987). Generations are continuous throughout the year.

EGGS

The small hemispherical eggs are about 1/25 inch (1 mm) in diameter and are colored yellowish green. When moth populations are low, a single female moth lays her eggs in batches of up to 100. When moth populations are high, eggs are laid in batches of several hundred eggs by individual females. Eggs are generally laid on the underside of leaves but may be found on the bark or on other plants nearby. Eggs hatch in 3 to 4 days at temperatures between 82.4^0 to 91.4^0 F (28^0 to 33^0 C) (Kumar and Lal, 1983).

LARVAE

This moth has 5 larval stages, or instars, separated by 4 molts. After each molt, the discarded skins are eaten by the newly emerged caterpillars. Caterpillars are cylindrical in shape and are 1/5 to 1/3 inch long during the first larval stage (Hargreaves, 1936). They reach up to two inches in length when fully grown (Tryon, 1898). Like other caterpillars belonging to the Noctuidae family, they have eight pairs of legs - three pairs under the thorax, four pairs in the middle portion of the abdomen (the first of which is rudimentary), and the last pair at the end of the abdomen. The last segment (the 11th) is considerably humped (Tryon, 1898), a feature that develops during the second instar (Hargreaves, 1936). They are either dark green to black or pale green to yellow (Comstock, 1936, Hargreaves, 1963). The dark coloration occurs when larval densities are high, and the light colored larvae are found with isolated larvae (Waterhouse and Norris, 1987). On the second and third abdominal segments, there are paired, lateral markings resembling eyes (Waterhouse and Norris, 1987). On the upper surface of the body, they have numerous small creamy-white spots and bars edged with black that tend to coalesce in some places (Tryon, 1898). Duration of the five larval stages at temperatures between 82.4^0 to 91.4^0 F (28^0 to 33^0 C) are 3 to 5 days, 3 to 5 days, 2 to 3 days, 2 to 5 days, and 4 to 10 days, respectively (Kumar and Lal, 1983).

PUPAE

Mature larvae pupate within a cocoon spun between leaves and woven together with silk. The leaves containing the cocoon may remain on the host plant or dry and fall to the ground (Waterhouse and Norris, 1987). Pupae are very dark brown with a purplish cast (Tryon, 1898) and about 1-1/8 inch (28 mm) long (Comstock, 1963). Pupation lasts for 14 to 21 days. If pupation occurs under very dry conditions the adult may not be able to emerge successfully (Hargreaves, 1936).

ADULTS

The adult moth is large and robust. It has a wingspan of almost 4 inches (10 cm) and a stout body, about 2 inches (5 cm) long (Waterhouse and Norris, 1987) that does not extend, or slightly extends, beyond the hindwings (Tryon, 1898). The eyes are large. The area behind the head of the moth, the thorax, is pale to purple-brown and the abdomen is pale brown at the base brightening to yellow-orange at the tip. The forewings resemble a leaf by being olive to purple-brown and may have white and green colored flecks (the colored flecks are more common on females). This leaf-like appearance of the forewings makes this moth difficult to see when it is at rest, especially, because the bright hindwings are not visible. The outer edges of the female's forewings are scalloped or toothed where those of the male's are evenly curved. The hindwings are bright orange, have a black comma-shaped mark and are fringed by a black border with white dots.

After emerging from the pupa, females have a preoviposition period of 4 to 8 days before she begins to lay her eggs. Each female may lay up to 750 eggs during her lifetime. Females live for 27 to 30 days and males 26 to 28 days (Kumar and Lal, 1983).

BEHAVIOR

Caterpillars mostly feed between 5:00 PM and 10:00 AM, but may feed at any time. They are usually located beneath or on the edges of leaves (Hargreaves, 1936). Young larvae drop to the ground at any sign of danger, while the older larvae take an aggressive attitude by hanging on to the food plant with their hind legs and swaying the rest of their body from side to side (Mosse-Robinson, 1968).

Although their flight is slow and heavy (Baptist, 1944), the adult moths are very strong fliers and can travel great distances from their breeding grounds in search of food (Waterhouse and Norris, 1987). Adults fly mainly between the hours of dusk and 11:00 PM (Tryon, 1898). They are not usually attracted to light (Kumar and Lal, 1983). When disturbed, the moth flares its forewings, exposing its conspicuous hindwings (Waterhouse and Norris, 1987).

MANAGEMENT

Biological Control

The low incidence of damage by this pest is attributed to the effectiveness of its natural enemies throughout most of the world. Two of the seven recorded parasites of the Pacific fruit-piercing moth, *Trichogramma chilonis* (Trichogrammatidae) and *Euplectrus plathypenae* (Eulophidae), are present in Hawaii. Five predators and a fungus are also reported in the literature, but presently, they have not been found in Hawaii. The parasites in Hawaii usually control this pest efficiently (Heu *et al.*, 1985).

Other Non-Chemical Control Measures

Several means of controlling fruit-piercing moths were evaluated in India (Baptist, 1944) and are outlined below.

Capture and Destruction of Moths :— Since this moth is not easily disturbed from fruit once it has begun to feed, netting and killing of moths is possible. This is best accomplished an hour after sunset when there is sufficient darkness with the aid of torches or a strong flashlight. This method is most feasible when fruit are easily accessible and populations of this moth are small. This method is not very effective once a large population of this moth exists and is usually recommended only if no other control means is possible.

Lanterns and Lights: — This moth prefers darkness and avoids light, therefore the illumination of orchards was tested as a possible means of deterring Pacific fruit-piercing moth attack. Although damage was less severe under illuminated conditions, this method was determined to be an impractical means of reducing damage based on a small scale trial (Baptist, 1944). However, Nomura (1965) reported a substantial reduction of moths by 60% in Japan and other investigators have also found this method effective (Whitehead and Rust, 1967, 1972; Bosch, 1971a; BŠnziger, 1982; Kumar and Lal, 1983). These moths are attracted by ultraviolet light and repelled by white light (Bosch, 1971a). If exposed to green-yellow light of mercury lamps, they adapt to the light and assume their resting daylight behavior (Bosch, 1971a).

Bagging or Screening Fruits :— Mechanical protection of fruit against moth attack was achieved by covering the fruit with brown paper or transparent oil paper bags. Brown paper bags last for about a month in the field and the transparent oil paper bags may be used for up to two seasons if the weather is not very wet. This method is most practical when each individual fruit is of significant value or when fruits are easily accessible and are of large size or relatively compact bunches. Although this method is labor intensive, it is especially good if regular attention cannot be given to the crop or it is desired to have the fruits fully ripen on the trees. This method is not practical when fruits are small, scattered, and not within easy reach.

Smoking of the Orchard:— This method involves the obscuring of the odor of mature and ripening fruit that attracts the moth with the odor of the smoke. Containers full of inflammable material, oil, tar and some green plant trimmings to enhance the smoke were placed within the orchard at a rate of 2 to 4 per acre. The smoking process was started a half an hour before dusk and continued for 2 to 3 hours after nightfall. This period represents the time in which the moths are seeking their nighttime feeding grounds. If the smell of the orchard is masked, the moths choose various wild hosts and remain on them throughout the night to feed. This method reduced some damage caused by the Pacific fruit-piercing moth in India (Baptist, 1944), but for the most part, it was ineffective in Thailand (BŠnziger, 1982).

Orchard Sanitation: — This method involves the regular collection and proper disposal of all attacked and spoiled fruit. Both fallen fruit and attacked fruit on the tree should be collected, buried deeply or boiled in water for 10 minutes then broken up for compost. These procedures dissipate the odor emanating from the spoiled fruit so they cannot serve as an attractant for the moths (Baptist, 1944). This method was somewhat effective in India (Baptist, 1944) but ineffective in

Thailand (BŠnziger, 1982). The ineffectiveness of this method in Thailand was attributed to the preference for fruit still on the tree by the moths (BŠnziger, 1982).

Forced Harvesting:— When severe infestations are anticipated, damage may be avoided by harvesting the entire crop as soon as signs of fruit ripening is observed. The grower must determine if the revenue loss from premature harvesting is less than that from possible damage (Baptist, 1944).

CHEMICAL CONTROL

The effectiveness of chemical baits is variable. In Thailand, BŠnziger (1982) reported chemical baits were ineffective, while Kumar and Lal (1983) reported satisfactory control in India. The ineffectiveness of baits is primarily based on the lack of the appropriate bait, but this method may become useful in the future as new baits are discovered (BŠnziger, 1982).

Repellents can be either taste repellents or odor repellents. Since odor is the initial attractant for getting the moths to the host sight, control methods using repellent sprays focus on odor repellents (Bosch, 1971b). Although Bosch (1971b) reported some success using citronella oil as a repellent, repellent sprays are generally considered ineffective (Baptist, 1944; BŠnziger, 1982).

Zolfagharieh *et al.,* (2006) took laboratory investigation on inherited sterility of pomegranate fruit moth (*Ectomyelois Ceratoniae* Zell) Pomegranate neck worm *Ectomyelois Ceratoniae* Zell is the most important pest in Iran. The average damage is 40% annually. Female ovipositing and larvae behaviour indicated that chemical control cannot be recommended, applying gamma ray as a genetic control is effective. The larvae reared in laboratory conditions (29±1.0°C, 65±5% RH, and 16 Hours light) 4 days pupal stage before eclosion was irradiated with 0, 70, 80, 90, 100, 200, and 300 GY doses. The adults investigated after eclosion treatments were Irradiated males x normal females, Irradiated females x Normal males, Irradiated male x Irradiated females. The eggs were collected, counted and eggs hatched was determined. The 0, 70, 80 GY had no significant difference and eggs hatched but at 90, 100, 200, 300 GY the eggs did not hatch. The best dose proposed was 90 GY.

Ozturk and Ulusoy (2006) studied the pest and natural enemies determined in pomegranate orchards in Turkey. It was found that 100 pests belonging to 12 orders (Insecta =8, Gastropoda, Arachnida, Aves, Mammalia) and 38 families and 64 natural enemies belonging to 7 order (Insecta=7) and 14 families in pomegranate orchard in Turkey. Depending on the region two pest among them *Ectomyelois ceratonia* and *Ceratitis capitata* are found to be the main pest and Aphis punicae, *Siphoninus phillyreae, Aleurothrixus flacosus, Planocoecus citri, Zeuzera pyrina, Carphophillus bipustulatus, Schistocerus bimaculatus,* are found to be other important pest that are problem today. As per natural enemies *Chrysoperla carnea, Coccinella septumpantata, Exochomus quadripustulatus, Serangium montazerii, Scymnus apetzi, Nabis punctatus Orius niger, Deraeocoris serenus, Encarsia inaron, Episyrphus bactaetus, Forticula auricularia,* are found to be common species in pomegranate orchards in Turkey.

Rahim and Mahamudul Huq (2006) studied crop management for controlling pomegranate fruit borer and powdery mildew. In Bangladesh Fruit borer and powdery mildew are the two major problems in pomegranate production. Integrated crop management like sanitation pruning, light trap, smoking, fruit wrapping and using balance fertilizers and sex pheromones significantly reduced the infestation of fruit borer and powdery mildew.

Kolekar (2009) studied the monitoring of the pesticide residue in exportable pomegranate fruits from different states of India. All the samples were monitored for 86 pesticides residues which included pesticides from diversified chemical classes including organochlorine, organophosphates, carbamates, synthetic pyrethroids, neoniconoids, trizines, trizoles, natural product derivatives and other using established and validated methods. Overall 35.5% of the total analysed samples contained no detectable residues 44.44% samples contained detectable residue with 20% samples contained the residue that exceeded the maximum residue limit (MRLs) set by European union. Majority of samples contained the residues from organochlorine, organophosphates, neonicotinoids and trizole group of pesticide. The result of the current study showed that no restricted or banned pesticides such as DDT – HCH and their isomers were found in any of the samples analysed. The violative percentage for the samples analysed during 2007-08 was 3.08% which increased 29.56% in the year (2008-09) indicating increase use of pesticides to control pest and diseases on pomegranate. The violative percentage increased dramatically due to presence of fungicide residue in 61.1% samples of the non confirmed samples.

Bagale, (2009) studied the varietal reaction, extent of damage and management of anar butterfly *Deudorix* (Virachola) *isocrates* in Pomegranate. There is a gradual increase in area of pomegranate and so also the pest problems like *Deudorix* (Virachola) *isocrates* in pomegranate which assumed a major status limiting the cultivation of the crop, an attempt was made to study varietal reaction, extent of damage and management to borer in commercially grown varieties of pomegranate, namely Ganesh, Mirdula, Joyti, Ruby and Jalore seedless during Hastabahar and early ambe bahar, need based irrigation was given at regular interval. Bagging was resorted for the management of borer, observation reveals that all the varieties are susceptible to borer attack. The incidence ranged between 35% in Jalore seedless to 42% in Ganesh. Gradually it increased from 50 to 60% in second week of March. The situation was deteriorated and almost 80-90%

fruits were found to be damaged by borer in unprotected plot. Bagging of fruits showed significant reduction on percentage damaged fruit and the extent of damage was observed to be below 10% in almost all varieties as against 90% in open fruits.

Zolfagharieh et al., (2009) evaluated the potential of sterile insect technique in controlling *Ecomyelois ceratoniae* in pomegranate in Iran. According to thermal, humidity conditions and varieties losses vary annually. The application of pesticide is not practical and other methods like biological, mechanical, agronomical alone can be effective. Application of sterile technique is a method that is used for a few insects under special condition under laboratory. Pest larval stage were collected from the Pomegranate orchards in Saveh, Iran and reared in laboratory conditions (28 ± 2°C 60 ± 5% RH), 14 Hours light) on artificial diet. Pupa at 1 – 2, 3 – 4 pupas were separately irradiated at dose range 0-1.6GY. The adult after eclosion were crossed according to the following situations (irradiated males and natural females and irradiated males x irradiated females) under 4 replications. After counting the laid eggs and determining the egg hatching the results revealed that, the best formulae was crossing irradiated males x irradiated females and the best controlling dose of 1 – 2 days pupa was 120 GY and 3 – 4 days was 160 GY.

Krishna Naik *et. al.,* (2009) studied the beetle enemies of pomegranate and their management. Beetle insect pest infesting pomegranate includes shot hole borer, *Xyleborus perforans* and cerambycid stem border, *Celosterna scrabrator*. The shot hole adult female bore in to basal part of the stem and root cause drying in pomegranate. Due to this death of plants varies from 8.30 to 46.3%. These pests can be managed by drenching with chloropyriphos 20 EC (4 ml/liter) with carbendizim (2 g/liter) or propiconozole 25 EC. The stem borer infections vary from 5 to 37%. The grub bore into stem and make deep tunnels (up to 60 cm) resulting in yellowing and drying of branches and even entire plants. Adults insects also scrape the growing shoots lead to drying. This pest can be managed by injecting 8% dichlorovas 76 SC (80 ml/liter) with squeeze bottle till the tunnel is filled and closed with wet mud.

Schlomo Marco, (2009) studied plant protection in high yielding pomegranate plantations present situation and prospectus for the future. In high yielding pomegranate plantation, plant protection is at present based mainly on chemical control. The inputs needed here for a sound plant protection regime are higher per area, as the plantations are aimed at higher yields and quality. As a consequence of 30 – 50% of the total running expenses of such conventional orchards are dedicated to plant protection activities. The higher the value of possible losses, higher is the pressure on the farmers 'not to miss' a protective treatment. This leads inevitable a will to reduce the number of treatments. Besides the obvious economical consideration, some ecological, epidemiological and practical, marketing problems evolve. Already now, serious restrictions on chemical use are imposed either by regulation or by the market, restrictions which are expected in the future to limit more and even cripple this sort of plant protection. The obvious short cut in the field domain is to maximize the efficiency of the present system, utilizing chemicals only when needed and can exert their maximal efficiency. This in turn will reduce number of treatments and quantity of chemicals in use. In order to achieve this efficiency we have to have detail information of the fluctuating situation in the field, population of pests, their interaction, epidemiology expected dispersion curve etc. It means that at least on regional scale we have to establish a reliable monitoring system in the fields formulated by experimental professional staff to decide the 'threshold level' that will launch the treatments. According to the experience this will reduce significantly the number of treatments and will enable a better choice of the proper materials in 'windows of opportunity' consideration. For the long run we should consider organic production which eliminates completely the chemical control. This is very important for the juice and pharmaceutical industrial products from pomegranate. Unfortunately high yielding plantation finds this system very difficult to achieve from the professional point of view. The organic pomegranates we find at present in the market are mainly the products of small farms or home gardens which do not spray chemicals mainly because they cannot afford to buy them. These products are the result of various organic methods like (sustainable, natural, green etc). This sort of products has many disadvantages but for our point of view, this is not a real organic product because it cannot full fill the vigorous and very stringent demand from an organic product. At the moment the most promising and practical approach will be to adopt the IPM (Integrated Pest Management) system of plant protection. The backbone of this approach is to integrate many other control possibilities (genetic, biological, cultural, agro-technical, physical, epidemiological) within the chemical control which in this way should diminish the pest problem.

Balikai *et. al.*, (2009) reported the status of pomegranate pest and their management strategies in India through scanning of literature revealed a total of 91 insects, 6 mites and one snail pest feeding on pomegranate crop in India. Out of these the most obnoxious enemy in pomegranate butterfly, *Deudorix* (=Virachola) *isocrates* Fab, which may destroy more than 50 percent of fruits. Next in order are three species of fruit sucking moth (*Eudocima* (=Othreis) *fullonia* (Linn) *Eudocima* (=Othreis) *maternal* Linn); *Eudocima* (=Othreis) *homaena* Hub) and two species of pomegranate shot hole borer (*Xyloborus fornicatus, Xyloborus perforans* Wollasan) have become major pest infesting in the collar region of the plant by making innumerable pin or shot holes causing discontinuity in the conducting vessels affecting conduction of water to the upper portion of the plant. Consequently the drying of the twigs became imminent. The drying progressed from the tip towards base and in severe cases entire plant dried up. Indiscriminate use of insecticide has led to many serious problems like white flies (pomegranate white fly *Siphoninus phillyreae*, spiraling white fly *Aleurodicus dispersus*), mealy bug *Pseudococcus lilacinus*, thrips (*Rhipiphorothrips Cruentatus*) *Scirthothrips dorsalis, Anaphothrips oligochaetus*, Aphid (*Aphis punicae* and mites (*Aceria granatia, Oligonychus punicae*. These sucking pests occurs during flowering and fruiting stage of the crop, thereby reduce the vigor of the plant in addition to excretion of honey dew on the leaves and development of sooty mould on leaves and fruits. Besides termites, *Odontotermes obesus* and root grub *Anomala dimitiata*, attach roots, defoliation feed on foliage and few bugs, borer, moths and fruit flies damage the fruits.

Kulkarni and Dethe, (2009) studied seasonal incidence of major pomegranate pest and population fluctuation were under taken. The incidence of insect pest was recorded at each meteorological week from July 2005 to June 2006 covering three pomegranate seasons called Mrigbahar, Hastabahar, Ambiabahar. The weekly meteorological data was also recorded during the period of observations. Occurrence of pest viz., fruit borer, thrips, aphids, white fly, mealy bug and scale insects was noticed, but the intensity of scale insects were negligible (0.01 to 0.49/twig). Relatively, Mrigbahar season was more favourable to infestation of fruit borer (10.82%), Thrips 4.36/fruit. However, incidence of aphids (39.58/shoot) in Hastabahar and mealy bugs (5.29/fruit) and White flies (4.72%) in Ambiabahar was recorded to maximum.

Anita Kumari *et al*,.(2009) studied population dynamics of major pests of pomegranate in central Telagana zone of Andhra Pradesh, India. India is the second largest producer of pomegranate with an area of 1.13 lakh hectare and production of 7.92 lakh tons. Bhagwa and Ganesh are the commercial varieties grown extensively in Andhra Pradesh. The major pests recorded were aphids, thrips, stem borer whitefly, fruit borer, mites and mealy bugs were recorded in reproductive parts. Five varieties of pomegranate viz., Arakata, Bhagwa, Mirdula, Ganesh G-137, were screened for pest incidence. Fruit borer *Deudorix isocrates*, is the major pest which is constantly and regularly injurious in pomegranate during flowering and fruit set stage. Oviposition was observed on Calyx and young larvae damage the fruits from fruit set stage to maturity. The incidence was maximum during monsoon season (Mrigbahar crop). Among the five varieties screened for fruit borer, maximum infestation was recorded in Mirdula (50%) followed by Arkata, G-137, Ganesh and lowest in Bhagawa (30%). Low temperature coupled with high humidity was congenial for the development of borer during second fortnight of August. Incidence of thrips was observed on new flush but its severity was less. Aphids infestation was noticed on new flush as well as small fruits and the infestation was severe during August to September months. Mites damage is also noticed but with less intensity. Mealy bug infestation was more in Hastabahar crop and increased with increase in temperature.

Wohlfarter *et al.,* (2009) reported *Nanophyes spp.* recorded for commercial pomegranate (*Punica granatum L*) in Western Cape Province of South Africa. Survey was conducted at a monthly interval for two seasons from September 2006 to May 2008. During September 2007 damage to young growing tips was recorded from Heidelberg in the Western Cape Province. Trees were inspected for insect presence and infested shoots were collected and placed in laboratory with approximate temperature of 26°C and relative humidity of 60% to rear adults. Additional adults were collected with help of a beating sheet and together with the emerged specimen and sent for identification, to Bio-systematics Division of the Plant Protection Research Institute, Agriculture Research Council (ARC-PPRI) in Pretoria, South Africa. Small legless and curved larvae which are characteristics of Nanophyids were found embedded in the succulent pomegranate shoots. According to Proche (2008) Nanophyids bored holes in to the tips of young shoots with their rostrum for egg deposition. The hatched larvae fed on soft plant tissue, hollowing out shoots and damaging young leaves. Weekened shoots usually break in the slightest breeze therefore stunting growth. Tree under heavy Nanophyid infestation as observed in single orchard near Picketberg, had a reduced flower load, therefore negatively affected production. No fruit damage was recorded. The ARC – PPRI confirmed the reared and collected curculioninae to be *Nanophyes spp.* Yet positive species identification could not be determined.

Kulkarni *et. al.,* (2009) studied the different insecticide efficacy for the control of *Deudorix isocrates* and their residue in pomegranate, which causes 40 to 90 percent damage, which is severe in rainy season from August to October when the temperature is moderately coupled with high humidity. Different control measures like bagging of fruit with paper/muslin/polyethylene bags. Spraying of carbaryl 50 WP at 0.2% or phenthoate at 0.05% are suggested. Residue of pesticide were studied in the arils of pomegranate fruits harvested periodically after third spray using the chromatography method (GLC – HPLC) at the detectable limit of 0.02 ppm for spinosad, deltamethrin and 0.05 ppm for emamectin benzoate, and thiodicarb. Spray treatment of 0.018% spinosad, 0.0022% emamectin benzoate, 0.15% neem seed extract were evaluated on the basis of percent fruit borer damage, fruit yield and economics of treatments. Each treatment consists of three spray applied at an interval of about three weeks starting from fruit set. Spray treatment of emamectin benzoate and spinosad managed the fruit damage to meager 2.37 and 3.21% and recorded the yield of 11.10 and 10.93 tonne per hectare respectively as compared to 26.33% fruit damage and 6.28 tonne/hectare yield in untreated plants. Net gain per hectare was maximum of Rs.88,495 in spray treatment of emamectin benzoate followed by spinosad (Rs.85,060), deltamethrin (Rs.48,532) and thiodicarb (Rs.47,100). Deltamethrin spray treatment adversely affected abundance of predators and egg parasitoids. Residues were below detectable limit (BDL) in fruits from plants sprayed with spinosad (0.018 and 0.036%), emamectin benzoate (0.0022 and 0.0044%) and deltamethrin (0.002%) against fruit borer. However detected residues of thiodicarb in 0-day samples (0.18 to 0.35 ppm) reached BDL of 0.05 ppm in 1.58 to 3.82 days.

Kulkarni *et. al.*, (2009) studied the efficacy of different insecticide for the control of thrips *Scirtothrips dorsalis* Hood and their residues in pomegranate. Spray treatments of 0.018% spinosad, 0.01% chlorfenapyr, 0.005% thiomethoxam, 0.005% acetamiprid, 0.015% Cypermethrin and 5% Neem seed extract were compared. Each spray treatment consists of three sprays with an interval of 15 days by initiating at fruit set. Residues pesticides were studied in the arils of pomegranate fruit periodically harvested after third spray using chromatographic methods (GLC and HPLC) at the detectable limit of 0.02 ppm for spinosad and cypermethrin and 0.05 ppm for chlorfenapyr and thiomethaxom. All pesticide were significantly effective in keeping the count of the thrips (No/fruit) below at 5, and 10 days after spray when compared to the count of 17.82 to 20.85 on untreated plants. As against fruit yield of 6.64 tonne/hectare in untreated plants. All pesticide treatment except Neem seed extract were significantly superior on par with each other. Adverse effect on predators was noticed in spray treatment of cypermethrin. Spray treatment evaluated against thrips viz., spinosad (0.018 to 0.036%). Chlorfenapyr (0.01 – 0.02%), thiomethoxam (0.005 and 0.01%) and cypermethrin (0.015%) were studied for residues of only thiomethoxam were found in 0 – day sample (0.08 – 0.19 ppm) which reached BDL of 0.05 ppm in 1.0 to 2.97 days.

Sreedevi and Abraham Vergees (2009) studied the effect of Aphid *Aphis punicae* passerine (Hemiptera : Aphididae) infestation on pomegranate flowers, fruit size, growth and development. Forty flower buds with aphid infestation were randomly selected and tagged in the field. Similarly, the extent of bud dropping, the percent aphid coverage and the fruit size (after fruit set) were recorded at different stages of bud development viz., bud, flower and fruit at 3-4 day interval till fruit matured. The fruit size was recorded by measuring the circumference of the fruit at each observation using a thread and measuring along a centimeter ruler. The buds without infestation were maintained as control throughout by removing the aphids, if any colonized in due course of the study. The documented data were subjected to correlation and regression analysis. It was observed that aphid infestation resulted in significant flower drop to nearly 84% of infested flower and young fruits as compared to 20% drop in uninfestated ones. In survived fruits, the size was negatively correlated with percent aphid coverage ($r = -0.790$). It necessitates the initiation of control measures at flower bud formation stage. The pomegranate fruit size was regressed against the percent aphid coverage on fruits to know the best model that could explain the relationship. The linear (y=0.299x + 20.882) and exponential models ($y = 20.987e^{-0.0202x}$) could explain this variability in fruit size due to percent aphid coverage to the extent of 89 and 88% respectively. The polynomial model ($y = -0.0037x^2 - 0.1735 + 20.784$) showed a CD of 90% explaining the reliability of variation to that extent.

Kulkarni *et. al.,* (2009) conducted a study on organic management of the pest using entomopathogenic fungus *Verticillium lecanii* (phule bugicide ®). An experiment in a randomized block design with four replications was conducted for four years. Five different concentrations of *Verticillium lecanii* (phule Bugicide ® WP) were tested against the pest. The observations were taken on live and dead insects with the help of hand lens. The mortality of insects at 3rd, 7th and 10th days after mortality was worked out. The pooled results indicated that the infestation of mealy bugs was in the range of 30.81 to 31.84 nymphs/fruit before application of spray. All the treatments recorded significantly higher population reduction of mealy bugs over untreated control at 10 days after spray. The treatment with *Verticillium lecanii* (6×10^5 cfu/g) @ 6g/lit was found superior in which the maximum (79.98%) population reduction of mealy bugs was recorded on 10 days after spray and the yield obtained was 19.98 (t/ha) with B:C ratio of 1:479

Khajuria (2009), pointed out that Anar butter fly, *Deudorix epijarbas* (Moore) are major production constrain in pomegranate cultivation in Kullu district of Himachal Pradesh. Presently, sprays with highly persistent insecticides like monocrotophos and cypermethrin are being used for its management. As a result of secondary pests like pomegranate mite, *Tenuipalpus granati* Sayed (Acrina: Tenuipalpidae) has been observed regularly in many orchards since last 5-6 years. A survey was conducted to know the nature and extent of damage caused by mites in different pomegranate area by making regular units. Pesticides application records were also checked in highly infested orchards in some of the localities.

A preliminary study on the biology was conducted on four marked trees at Seabagh. Four selective miticides viz., dicofol (0.03%) fenpyroximate (0.0025%), propergite (0.057%) and Fenanzaquin (0.0025%) were evaluated by spraying at sizeable population of the mite at Bajaura during 2005 for generating adhoc recommendations to the farmers. Studies revealed that the mite species. *Tenuopalpus granati* has attained a pest status in all orchards of the valley. Hurla, Nagwain, Bajaura and Jia villages were the top spots with average population of mites ranging from 80-145 per leaf. This broad, red coloured, sluggish mite species fed on the leaves and removed chlorophyll resulting in loss of green colour. Such leaves first turned dull whitish yellow and then bronzed and finally dried and defoliation was observed. The fruit remained under sized in such trees and entire crop in an infested orchard was lost. Studies further revealed that the mite appeared in those orchards where only cypermethrin was sprayed repeatedly. The mite over wintered as orange coloured female on the trees from mid August to March. In April these over-wintering females started laying red coloured eggs in rows on the underside of the leaves. There were 2-3 generations in a season. All the four acaricides evaluated against the mite caused 96.0 to 99.2% mortality. Only one spray provided good control to the mite populations under field conditions.

Kotikal *et. al.,* (2009), they conducted an experiment to know the incidence of major sucking pests of pomegranate and their relation with weather parameters. Fixed plot survey conducted at five locations revealed the presence of pomegranate aphid (*Aphis punicae* passerine), thrips (*Rhipiphorothrips cruentatus* Hood; *Scirtothrips dorsalis* Hood; Anaphothrips oligochaetus kerny), white flies (Ash whitefly, *Siphoninus phillyreae* Holiday; sporalling whitefly, *Aleurodicus disperses* Russel), mealy bug (*Planococcus lilacinus* (cockrell) and mites (Eriophid mite, *Aceria granati canestrini*; False spider mite, *Tenuipalpus granati* sayed). A high activity of aphids was observed during second fortnight of December onwards which reached peak during first fortnight of January. Simple correlations between incidence of aphids and weather parameters revealed that only morning relative humidity showed positive and significant relationship, which may lead to increase in percent hatching of eggs, whereas all other weather parameters showed negative and non-significant relationship with incidence of aphids. Among the various antibiotic factors minimum temperature and evening relative humidity influenced aphids to the extent of 67 and 32 percent, respectively. The peak infestation of thrips was noticed during second fortnight of March. This was due to the fact that tender leaves and fruits were available during the respective periods which favoured the multiplication of the pest. Simple correlations revealed that only morning relative humidity had significant and negative relationship with thrips incidence, but remaining weather parameters recorded non-significant relationship with thrips. Population density of whiteflies was maximum from second fortnight of February to first fortnight of March. Only morning relative humidity showed positive and significant relationship with white fly incidence, while evening relative humidity, rainfall, number of rainy days also showed significant and negative relationship. Nymphs and adults of mealy bug appeared during early February and caused maximum damage during March and April months which coincided with the fruiting stage. The correlation studies revealed that morning relative humidity had significant and positive relationship whereas, evening relative humidity had significant negative relationship with mealy bug and influenced to the tune of 35 and 33 percent, respectively. Incidence of mites started from January and reached a peak during first fortnight of March. Simple correlations showed that increase in morning relative humidity favoured the multiplication of mite pest. Evening relative humidity, rainfall and number of rainy days had negative and significant relationship with mite population and these factors reduced the incidence of mite. Among all abiotic factors, evening and morning relative humidity influenced mite population to the extent of 53 and 36 percent respectively. To develop pest forecast models, correlation studies were made between weather parameters at two weeks lead time and population of sucking pests. Morning relative humidity recorded positive and significant correlation while evening relative humidity, number of rainy days, recorded significant and negative relationship with all sucking pests and mites at two weeks lead time. Accordingly, if the morning relative humidity increases then the population of sucking pests could be higher during next two weeks.

PESTICIDE CONTROL SYSTEM

Most countries have brought out legislation and maximum residue limits have been specified for various horticultural produce including pomegranate. For ensuring pesticide control, it is essential to have an effective control system and assess the effectiveness of the same. The following measures are essential for having an effective pesticide residue control system:

1. A pesticide usage programme of the producer shall be submitted prior to production for each season.
2. A field pesticide application record shall be maintained by the producer.
3. Pesticide residue limits as specified by the importing country shall be maintained. Pomegranates for export to a certain country should be tested for limits specified by that country. All consignments of pomegranates for exports to the European Union shall be compulsorily tested for pesticide residues. These tests shall be conducted for those chemicals which are either banned or where there is a possibility of having higher residue levels than those permitted. A list of chemicals requiring compulsory testing for export to the European Union.
4. Routine random pesticide residue analysis shall be carried out as per the guidelines given in Guide to Codex Recommendations Concerning Pesticide Residues; Part 5 Recommended Method of Sampling for the Determination of Pesticide Residues.
5. In case of high residue levels, a plan needs to be drawn up to investigate into the cause and take corrective action to prevent future occurrences. This should include:
    a. An investigation into the possible causes of high residue levels.
    b. Re-sampling or repeat of the residue analysis with, if possible, an additional independent analysis.
    c. Withdrawal from sale or distribution of product from which the sample was taken and its isolation from any other product until the product has

been declared safe.

d. Informing the appropriate authorities and customers of the high residue levels as soon as the same is detected if the product is in the marketing chain.

6. The international, European and Indian limits for pesticide residues as given in the following publications.

International

- Codex Alimentarios Volume 2 and 2 B (1996) Pesticide Residues in Food (Second Edition), 2000.
- Codex Alimentarius Volume 14 Acceptances

European

- Council Directive 76/895/EEC of 23rd November, 1976 relating to the fixing of maximum levels for pesticide residues in and on fruit and vegetables.
- Council Directive 90/642/EEC of 27th November, 1990 on the fixing of maximum levels for pesticides residues in and on certain products of plant origins, including fruit and vegetables.
- Council Directive 93/58/EEC of 29th June, 1993 amending Annex 11 to Directive 76/895/EEC relating to the fixing of maximum levels for pesticide residues in and on fruit and vegetables and the Annex to Directive 90/642/EEC relating to the fixing of maximum levels for pesticide residues in an on certain products of plant origin, including fruit and vegetables, and providing for the establishment of a first list of maximum levels.
- Council Directive 94/30/EC of 23rd June, 1994 amending Annex 11 and Directive 90/642/EEC relating to the fixing of maximum levels for pesticide residues in and on certain products of plant origin, including fruit and vegetables and providing for the establishment of a list of maximum levels.
- Council Directive 96/32/EC of 21st May, 1996 amending Annex 11 to Directive 76/895/EEC relating to fixing of maximum levels of pesticide residues in and on fruit and vegetables and Annex 11 to Directive 90/642/EEC relating to the fixing of maximum levels of pesticide residues in and on certain products of plant origin, including fruit and vegetables and providing for the establishment of a list of maximum levels.
- Council Directive 97/41/EC of 25th June, 1997 amending Directives 76/895/EEX, 86/362/EEC, 86/363/EEC and 90/642/EEC relating to the fixing of maximum levels for pesticide residues in and on, cereals, foodstuffs of animal origin and certain products of plant origin, including fruits and vegetables.

- Council directive 2000/24/EC of 28th April, 2000 amending the annexes to Council Directives 76/895/EEC, 86/362/EEC, 86/363/EEC and90/642/EEC on fixing the maximum levels of pesticide residues in and on cereals, foodstuffs of animal origin and certain products of plant origin including fruits and vegetables.

Indian

- Fruits and Vegetable Grading and Marking Rules, 2002.
- Prevention of Food Adulteration Act, 1954 and Rules Revised 1998 framed there under.

CHEMICALS REQUIRING COMPULSORY TESTING FOR EXPORT TO EUROPEAN UNION

A MERCURY COMPOUNDS

1. Mercuric oxide
2. Mercurous chloride (calomel)
3. Other inorganic mercury compounds
4. alkyl mercury compounds
5. Aloxylalkyl + aryl mercury compounds

B. PERSISTENT ORGANO-CHLORINE COMPOUNDS

1. Aldrin
2. Chlordane
3. Dieldrin
4. DDT
5. Endrin
6. HCH containing less than 99 % of the gamma isomer
7. Heptachlor
8. Hexachlorobenzene
9. Camphechlor

C. OTHER COMPOUNDS

1. Banned substances:
   a. Ethylene oxide
   b. Nitrofen
   c. 1,2-dibromoethane
   d. 1,2-dichloroethane
   e. Dinoseb (acetates and salts)

f. Binapacryl

g. Captafol

2. The substances below are not banned if they meet purity requirements stated against each:

a. Dicofol — Products containing more than 78% of p.p. dicofol and less than 1 g/kg DDT related compounds.

b. Maleci hydrazide — Choline, potassium and sodium salts containing less than 1 mg/kg free hydrazine (acid equivalent)

c. Quintozene — Products containing less than 1 g/kg HCB and less than 10 g/ka pentachlorobenzene.

MAXIMUM PERMITTED PESTICIDE RESIDUE LEVELS FOR POMEGRANATE FRUITS-EC, CODEX AND INDIAN

Pesticides/Chemicals	Limits (mg/kg)		
	EC	CODEX	Indian
Acephate	0.02		
Aldrin and Dieldrin			0.10
Aminotriazole	0.05		
Aramite	0.01		
Atrazine	0.10		
Azenphos methyl		1.00	
Barban	0.05		
Benafurocarb	0.05		
Benalaxyl	0.05		
Benomyl			5.00
Binapacryl	0.05		
Bromide ion		20.00	30.00
Bromophos ethyl	0.05		
Camphechlor	0.10		
Captafol	0.02		
Captan			15.00
Carbendazim			5.00
Carbofuran	0.10		0.10
Carbosulfan	0.05		
Chlorbenside	0.01		
Chlorbufam	0.05		
Chlordane		0.02 *	0.10

Chlorfenson	0.01		
Chlorobenzilate	0.02		1.00
Chloropyrifos			0.50
Chloropyrifos methyl	0.05		
Chlorothalonil	0.01		
Chloroxuron	0.05		
Copper oxychloride			20.00
Cyhalothrin – lambda	0.02		
Cyfluthrin	0.02		
Cypermethrin	0.05		
D.T.T.	0.05		3.50
Daminozide	0.02		
Diallate	0.05		
Diazinon	0.02		
Dichlorprop	0.05		
Dichlorvos			0.10
Dicofol			5.00
Dimethoate			2.00
Dinoseb	0.05		
Dioxathion	0.05		
Disulfoton	0.02		
Dithiocarbamates			3.00
Endosulfan		2.00	2.00
Endrin	0.01		
Ethion			2.00
Ethylene dibromide	0.01		
Fenarimol	0.02		
Fenchlorphos	0.01		
Fenitrothion			0.50
Fentin	0.05		
Fenvalerate	0.05		
Formathion			1.00
Furathiocarb	0.05		
Heptachlor	0.01		
Hexachlorocyclo hexane (all isomers)			1.00
Lindane			1.00
Malathion			4.00
Maleic hydrazide	1.00		

Maneb or Mancozeb	0.05		
Mecarbam	0.05		
Methamidophos	0.01		
Methoxychlor	0.01		
Methylbromide	0.05		
Monocrotophos			1.00
Paraquat	0.05		0.05
Parathion			0.50
Parathion methyl			0.20
Phorate	0.05		
Phosalone			5.00
Phosphamidon			0.20
Propyzamide	0.02		
Pyrethrins			1.00
TEPP	0.01		
Thiometon			0.50
Trichlorofen			0.10
Triforine	0.05		
Vinchlozolin	0.05		
2,4-D			2.00
2,4,5-T	0.05		

Chapter 13

Disease in Pomegranate

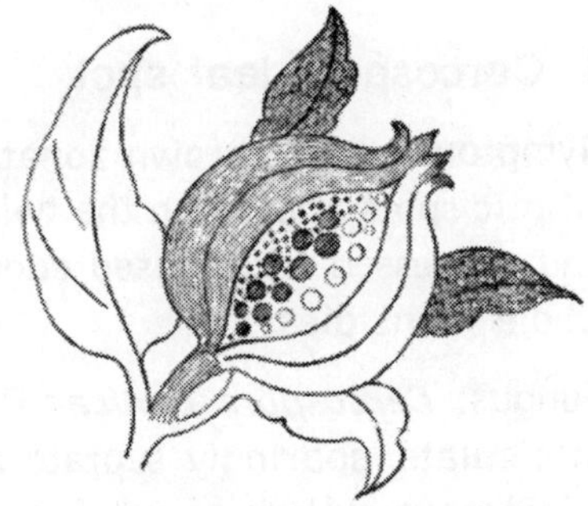

POMEGRANATE (*Punica granatum* L.)

1. Cercospora leaf spot - *Cercospora punicae*
2. Bacterial blight - *Xanthomonas axonopodis pv. Punicae*
3. Leaf spots - *Colletotrichum gloeosporioides*
   - *Sphaceloma punicae*
   - *Fusarium fusarioides*
   - *Phomopsis aucubicola*
   - *Drechslera rostrata*
4. Minor diseases
   - a. Canker - *Ceuthospora phyllosticta*
   - b. Leaf and fruit spot - *Coelophoma empetri*
   - c. Flower and fruit spot - *Phytophthora nicotianae*
   - d. Fruit spots - *Baltraniella humicola*
     - *Pestalotiopsis versicolor*
   - e. Fruit rots
     - i. Cladosporium fruit rot - *Cladosporium oxysporum*
     - ii. Aspergillus fruit rots - *Aspergillus spp.*
     - iii. Mild soft rot - *Penicillium chrysogenum*
     - iv. Soft rots - *Rhizopus arrhizus* & *R.stolonifer*
     - v. Dry rot - *Syncephalastrum racemosum*, *Coniella granati*
     - vi. Fusarium rot - *Fusarium equiseti*
     - vii. Phomopsis rot - *Phomopsis aucubicola*

5 Pomegrante wilt - *Ceratocystis fimbriata*

1. Cercospora leaf spot

Symptoms: Light brown zonate spots appear on the leaves and fruits. Black and elliptic spots appear on the twigs. The affected areas in the twigs become flattened and depressed with raised edge. Such infected twigs dry up. In severe cases the whole plant dies.

Fungus: *Cercospora punicae* P. Henn. Conidiophores are olivaceous brown, short, fasciculate, sparingly septate and 20 to 40 x 3 um. Conidia are hyaline to pale olivalceous cylindric, sub-fusoid to sub-clavate, septate and 40 to 50 x 3.0um.

Mode of spread and survival : The pathogen spreads through wind-borne conidia.

Epidemiology: The disease is serious during Sep-Nov.

Management: The disease can be effectively controlled by pruning and destruction of diseased twigs followed by spraying with Thiophinate-methyl 0.1 per cent or chlorothalonil 0.2 per cent or mancozeb 0.2 per cent.

2. Bacterial blight

Symptoms: Small irregular, water-soaked spots appear on the leaves. Spots vary from two to five mm in dia with necrotic centre of pin-head size. Spots are translucent, later turn light brown to dark brown and are surrounded by prominent water-soaked margins. Spots coalesce to form large patches. Severely infected leaves fall off. The bacterium attacks stems, branches and fruits also. On the stem, the disease starts as brown to black spots around the nodes. It leads to girdling and cracking of nodes. Finally the branches break down. Brown to black spots appears on the pericarp with L or Y shaped cracks. The spots on fruits are raised and oily in appearance.

Bacterium: *Xanthomonas axonopodis* pv. *Punicae* (Hingorani & Singh) Vauterin et al. (Syn. *Xanthomonas campestris* pv. *Punicae* (Hingorani & Singh) Dye). It is a Gram-negative rod, motile with single polar flagellum and 0.5 to 1.0 to 2.5 um. It is non-acid fast and aerobic.

Mode of spread and survival: The bacterium survives on the tree. The pathogen survives for 120 days on the fallen leaves during off season. The primary infection is through infected cuttings. The disease spreads through wind splashed rains, insects, secature.

Epidemiology: High temperature and low humidity favour the disease.

Management: Clean cultivation and strict sanitation in the orchard help to reduce the disease incidence. Spraying with Bordeaux mixture 1.0 per cent controls the disease.

3. Leaf spots

a. *Colletotrichum gloeosporioides* Penz. The disease appears as small, regular to irregular dull violet or black spots on the leaves. These spots are surrounded by yellow margins. The infected leaves turn yellow and drop off. The pathogen spreads through wind-borne conidia. The disease is severe during August-September when there is high humidity and the temperature between 20 and 27^0 C. The disease is effectively controlled by spraying with carbendazim 0.1 per cent or thiophanate-methyl 0.1 per cent or mancozeb 0.2 per cent at fortnightly interval.

b. *Sphaceloma punicae* Bitancourt & Jenkins. The disease attacks leaves, shoots, calyx and fruits. Rusty spots appear on leaves. Infected leaves turn yellow and die. Rusty coloured pustules appear on fruits. Drizzling rains and abundant dew favour disease development and spread. Spraying with thiophanate methyl 0.1 per cent or carbendazim 0.1 per cent controls the disease.

c. *Pestalotiopsis versicolor* (Speg.) Stey. The disease manifests its symptoms as minute, brown to rust coloured spots on the fruits. The spots coalesce with the disease advance and causes necrotic patches. The central portion of the lesion is depressed inward with raised margin and severe infection tear open the rind. In severe cases infection penetrates deep into the fruits and causes discolouration of seeds. Mycelium of the fungus is cottony white with yellowish tinge and branched. Conidia borne on short conidiophore. They are 5-celled, oblong to ovate, 21.8 x 4.3 μm in diameter. Intermediate three cells are dark coloured and long. Apical cells on either side are hyaline. Apical conidial cell bears 3 hyaline cells. The basal cell is short, obtuse, hyaline with a small pedicel.

d. Fruit rots:-

i. *Cladosporium oxysporum* Berk & Curt. The diseased fruits develop orange-red to dull-brown circular spots and become olive-brown. In advanced stage, the entire fruit rots. Hyphae of the fungus are septate, light olive green, 2.5 to 3.0 um in width. Conidiophores are light brown and simple. Conidia are light brown to olive green, 1-celled, fusoid and 1 to 20 x 3.5 to 4.5 μm.

ii. Aspergillus fruit rots: *Aspergillus flavus* Link. It causes brownish discolouration, which gradually becomes blackish and slimy. Later it gets slightly depressed and is covered by green conidial heads of the fungus. The disease causes soft rot of fruits and emits fermented odour.

Aspergillus niger. Symptoms first appear in the form of a brownish discolouration which gradually become blackish brown. In advanced stages, a major portion of the fruit is involved giving it a depressed and slimy appearance. Rotten fruits emit a fermented odour.

Aspergillus niveus Bloch. It produces small brownish patch which increases in diameter and turns darker with blackish tinge in the centre. The diseased tissue shrinks and ultimately disintegrates especially in the central region, followed by irregular depression and exudation of slimy mass. The diseased fruits emits foul odour.

Aspergillus versicolor. It produces small brownish patch which increases in size and turns darker with blackish tinge in the centre. The diseased tissue shrinks and ultimately disintegrates especially in the central region followed by irregular depression and exudation of slimy mass emitting foul odour.

iii. **Mild soft rot:** *Penicillium chrysogenum* Thom. Soft, watery spots of two to four cm diameter appear on the fruits. These spots enlarge and coalesce together. The spots are found covered with white mycelium and bluish green spores.

iv. **Soft rot:** *Rhizopus arrhizus* Fisher and *R. stolonifer* Ehr. Ex Fr. Small spots appear on fruits. They increase in size and coalesce. Infection is restricted to rind. But the entire internal content decays into a pulpy mass. Under dry conditions cracking originates from the point of infection. Packing straw should be treated with sulphur dioxide. Treatment of fruits with linseed oil or mustard oil or castor oil protects them from rot by *R. arrhizus.*

v. **Dry rot:** *Syncephalastrum racemosum* Cohn. Small isolated dark patches are formed on the surface of the fruits. The patches are dry and covered with the mycelium and spores of the fungus. Inner pulp rots.

vi. *Fusarium equiseti* (Corda) Sacc. Circular and depressed lesions appear on the fruits. Lesions are surrounded by concentric wrinkles. The lesions increase in size and cover almost the entire fruit.

vii. *Phomopsis sp.* The disease starts from calyx end and gradually spreads over the entire fruit. Pycnidia appear on affected areas. One spray with copper oxychloride checks the spread. Copper oxychloride 0.5% spray three times at 10 days interval controls the disease.

e. *Fusarium fusarioides* (Frag & Cif.) Booth. The disease appears as minute specks towards the leaf margin. The spots are brown, circular to irregular in shape. Later the spots coalesce and form big dark brown necrotic blotch.

d. *Phomopsis aucubicola* Grove. Buff-brown spots appear on the margins. Sometimes spots are formed scattered on the leaves. Black pycnidia are found on the spot in the upper surface of the leaf. Hyphae of the fungus are colourless, poorly branched, septate, 1.4 to 3.8 μm wide. Pycnidia are densely gregarious, black, immersed, globose to sub-globose, 201.6 to 489 μm in diameter, dark brown to black and ostiolate. Conidiophores are simple, short, unbranched, hyaline and 4.6 x 8.2 µm long. Alpha spores are hyaline, ovoid to oblong or rarely fusoid to sub-fusoid, often biguttulate and 7.4 to 11.8 x 2.4 to 3.1 µm. Beta spores are filiform, curved or unicinate and measure 19.8 to 34.4 x 0.54 to 1.4 µm.

e. *Drechslera rostrata* (Drech.) Richardson & Fraser. The disease is characterized by the appearance of numerous, small, black spots scattered all over the fruit. The margin of the spots varies from dark green to orange in colour. In advanced stages, few spots gradually enlarge and coalesce to form big dark spots of various sizes. Mild infection is confined to rind of the fruits but severe infection extends to the inner tissues and even to the seeds, showing ashy discoloration. The fungus produces conidia which are pale to dark olivaceous brown, cylindrical to rostrate, somewhat less curved, measuring 43.1 to 124.9 x 9.2 to 17.4 µm in size with 5 to 13 transverse septa and a minute hilum protruding from the base. The other end is bluntly tapered. The septa from basal and apical cells are prominent being thicker and darker than the intermediate septa.

4. Minor diseases:

a. **Canker:** *Ceuthospora phyllosticta* C. Mass. Elluiptic, black spots are formed on the twigs. Affected areas become flattened and depressed with raised edge. Later the bark dries and cracks and the wood below shows abnormal dark brown-black discolouration. Twigs beyond the cankerous spots dry off and severe cases the affected tree dies.

b. **Leaf and fruit spot:** *Coelophoma empetri* (Rostrup) **Petrak.** On leaves the spots are circular, reddish brown to dark brown. They coalesce to form bigger sized lesions which are necrotic and dark brown. Later infected leaves turn pale yellow and drop down. On fruits numerous, minute, circular, tan brown spots which turn brown to black later. They coalesce to form irregular, depressed and hard necrotic lesion. Lesions are restricted to epidermis bearing black, spherical pycnida. Mycelium of the fungus is olive-grey when fresh and it turned dark violet at maturity. They are separate, branched, and 5.5 µm in width. Pycnidia are dark brown to black, spherical to irregularly elongate and 335. 2 x 26.2 to 61.0 µm in diameter. Pycnidiospores are one septate, elongate to cylindrical with round to pointed ends and 11.8 x 1.6 µm.

c. **Flower and fruit spot:** *Phytophthora nicotianae* Breda de Haan. Spots on flowers leads to premature shedding. Lesions are also found on fruits. Twigs in the trees are also infected. Hyphae of the fungus are 4.8 µm in dia with hyphal swellings. Chalamydospores are intercalary or terminal. Sporanging are broadly turbinate with spherical basal portions and apical part prolonged into a beak, papillate and 63 x 44 µm. Sporangia do not shed.

d. Fruit spots:

i. *Baltraniella humicola.* The disease is characterized by appearance of black circular spots which gradually enlarge and coalesce to form big dark black spots leading to necrosis. The margins of spots vary from reddish to brown in colour. Infection is restricted to the rind of the fruit and underside of the pulp.

Gazel *et. al.,* (2006) reported detection of Hop stunt viroid in pomegranate trees in the East Mediterranean regions of Turkey. Hop stunt viroid (Hs Vd) infects several fruit tree species with agronomic importance including pomegranate. Different pomegranate cultivars growing in Hatay, Adana, Icel province were surveyed for the presence of Hs Vd. The main symptoms were yellowing and some alteration of leaves after analyzing 152 leaf samples for the presence of Hs Vd. Total RNAs were isolated from pomegranate leaves and used as template in RT – PCR. Hs Vd was not detected in all tested pomegranate.

Roman Labuda (2006) reported penicillium pathogen causing a destructive rot on pomegranate fruits *Penicillium implication.* Biourge was the cause of a destructive rot of pomegranate in Nitra, Slovak Republic. The ability of the *P.implication* strain to decay pomegranate under laboratory condition after inoculation with conidia in to subcutaneous layer of healthy fruits, which was seen on calyx, it is concluded that stamens are the source of infection from where it spread out following a mechanical damage during harvesting and storage or transport. This pathogen evoked a brown rot in some apple varieties (var. Jonagold, Selena, Vanda) there it seems that *P.implication* is another newly recognized potential plant pathogen from genus penicillium.

Ravikumar *et. al.,* 2009, studied the management of bacterial blight of pomegranate through chemical/antibiotics.

Bacterial blight of pomegranate *Xanthomonas axonopodis* Pv *punicae* has become an increasing serious threat to pomegranate growers in India. Spray with streptocyclin (500 ppm) + copper oxy chloride (2000 ppm) were found very effective in reducing the mean disease incidence (25.5%) followed by Bromopal (500 ppm) + copper oxy chloride (2000 ppm) (33.3%), when compared to control (78.5%) after 8th spray. The maximum mean yield of 9.3 Tonnes/ha recorded in streptocyclin (500 ppm) + copper oxy chloride (2000 ppm) followed by 8.5 ton/ in Bromopal (500 ppm) + copper oxy chloride (2000 ppm). The untreated check yielded 2.95 tons/ha.

Yenjerappa *et. al.,* 2009, studied the management of bacterial blight through cultural methods, like dates of pruning and rest period. Six different dates of pruning from September to November at fortnight intervals was observed for bacterial blight severity. Negligible bacterial blight (15.8%) and (13.8%) were reported in September I and II fortnight pruned crop. Moderate disease incidence was recorded in October pruned crop and more disease severity 60.7% was recorded in November II fortnight pruned crop owing to prevalance of congenial weather factors such as temperature, relative humidity and rainfall. Least disease index (6.59%) was recorded when the plants were given 6 months rest, and high disease incidence of 80.36% was recorded in one month rest crop. The reason being presence of heavy amount of inoculum carried from previous crop to the subsequent crop given one month rest. On the contrary inoculum building up was very less in the crop having longer rest period of six months, hence the lower disease incidence.

Somashekhar *et. al.,* 2009, studied the management of pomegranate wilt. A systematic survey was carried out in India which equates a loss of Rs. 3.43 crores during 1996 – 2005 with 5.69% of the pomegranate area affected with wilt [*Ceratocystis fimbriata*] above 5 years old plants affected with scolytid beetle (*Xyloborus fornicatus*) in the collar region (0.41%). All the leaves turn to yellow in colour in one branch sudden wilting of plant, epinasty symptoms, vertical stem cracking, blue stain patches on bark, brown streaks in the xylem, pin holes with discoloration. The same is reported from China also infected seedlings, irrigation water root contact, implement, rain water, budding, knife, secature, wind are the major source of spread. Scolytid beetle act as vector as it carry inoculum along with them. It survives in infected plants up to 190 days and in the soil for long time. Various chemicals were screened against *C. fimbriala* viz propiconizole, boric acid, phosphoric acid. Replanted plants grow normally after sterilization. The lines Masta, Bassein seedless, Cranado – de – Etcho showed resistant to wilt. Eradication of infected plant, soil sterilization with 2% Basamid (30 g/sqm) soil trenching between the plants. Soil drenching with propiconazole (0.1%), Boric acid (0.5%), Phosphoric acid (0.5%) to the surrounding plants or entire garden help in re-establishment of wilt affected garden.

Sharma and Teja (2009), reported the dry rot of pomegranate in Himachal Pradesh. Pomegranate is highly susceptible to dry fruit rot caused by *Coniella granati.* The rot causes softening of the rind, underneath pulp and seed. Pre harvest spray of carbendizim gave significant control. The Non-systemic fungicide like Mancozeb showed some promise. Fumigation with sodium or potassium metabisulphate was superior in checking the disease. Bavistin wrapper and Nipro fresh coating exhibited significant protection of pomegranate against dry fruit rot. Fungal antagonists viz., *Trichoderma virens, T. harzianum, T.hamatum, Trichoderma polysporium* gave maximum inhibition of mycelium growth of *Coniella granati* was obtained with citronella oil at 1000 ppm. Antagonist protection was higher when applied at a concentration of 10^8 spores/ml. Pre-inoculant treatment with plant extract of darek (*Melia azadirach*), dedonia (*Dedonia viscosa*) lanata (*Lantana camera*) and peppermint (*Mentha piperita*) provided significant control of fruit rot of pomegranate.

Yehya selahavarzi *et. al.,* (2009) reported about antifungal activity of aqueous and methanolic extract of different parts of pomegranate tree on some post harvest fungi using natural compounds such as plant extract and essential oil is an important method of controlling diseases. Different parts of pomegranate tree such as leaf, peel, bark have the inhibiting effect on fungal growth and spread. Therefore effects of methanol and aqueous extract of different pomegranate parts (peel, bark, leaves) were investigated on some fungi in the post harvest period of PDA medium. Different concentration of metabolic extract (0, 250, 500 750 il) and aqueous extract (0, 5, 10, 15%) were compared. The effects of these treatments were also studied on germination. Highest concentration of metabolic extract of pomegranate peel (750 il) had the most antifungal activity as growth of *Pencillium italicum. Rhizopus stlonifer, Aspergillus niger* fungi were inhibited to an extent of 100, 92 and 85% respectively. The extract also completely inhibited germination of *Pencillium italicum* and *Rhizopus stlonifer* spores. Only 9% germination was observed in *Aspergillus niger.*

Jamadar *et. al.,* (2009) reported about status of pomegranate diseases and their management in Northern Karnataka. The area under pomegranate is fast dwindling due to onslaught of number of diseases such as bacterial blight, anthracnose, wilt and dieback or decline. The bacterial blight of pomegranate (*Xanthomonas axonopodis pv. Punicae*) has assumed epidemic proportion (40-85% severity) in North Karnataka (Bijapur, Bagalkot, Bellary) seriously threatening its cultivation. On leaves small water soak lesion, while on fruit doty spots with L or Y shaped cracks are found. The pathogen survives in fallen leaves and fruits during off season and spread through planting material, cutting secature, wind, splashed rains. Temperature around 31.8 to 35.5°C with enough rainfall during July-October found to favour disease development. Pruning in September-October, hastabahar cultivation and rest period are highly beneficial to pomegranate plants and suppress the pathogen. Spraying Bacterinashak or streptocyclin (500 ppm) + copper oxy chloride (2000 ppm) has been found more effective. Pomegranate wilt *Ceratocystis fimbriata* is the most important disease in Karnataka and Maharashtra. Symptoms include yellowing of leaves, epinasty and sudden wilting. The infection spread through infected seedlings, sanitation measures such as removal destruction of affected trees, soil sterilization with Formalin (20 ml/liter) check the disease. Pomegranate decline also referred wilt involves poor soil and water management practices. Use of Black and heavy and ill drained soil, close planting, excessive irrigation and fertilizer pre-dispose the plants for dispose complex. *Rhizoctonia bataticola, Fusarium solani* and root knot nematode as well as shot hole borer. Management includes soil drenching with propiconazole (0.1%) + boric acid (0.5%) + phosphoric acid (0.5%). Anthracnose of pomegranate caused by *collectotrichum gloeosporiodes* is most serious during July – October. The disease appear as small, irregular to regular black spots on leaves. Calyx region and fruit which turn later on as dark brown depressed spots. Rain, high humidity and temperature of 20 – 27°C favours the disease, management through carbendizim, difenconozole or Thiophenate methyl @ 0.1% or chlorothalonil 0.2% spray at 15 days interval has been found effective. The minor disease are Alternaria blight, leaf spot due to cercospora, sphaceloma, Fusarium, phomopsis, Drechslera.

Phytophthora nicotiana and Canker. (*Ceuthospora phyllosticta)* as well as fruit spots and rots.

Benagi *et. al.,* (2009), surveyed pomegranate disease in northern Karnataka. The survey was carried out during Mrig and Hastabahar of 2008-09 in major pomegranate areas in North Karnataka state, Districts like Bagalkot, Bijapur, Belgaum, Bellary, Gadag and Koppal. Plantation with 3 year age selected. Five major and three minor diseases were observed, they were bacterial blight, wilt, anthracnose, Alternaria, Root knot nematode, scab, Fruit rot. Assessment of bacterial blight was made based on Leaf infection, stem canker, fruit infection. Finally severity of the disease on tree was calculated. Bacterial blight is in severe form in all pomegranate area which ranged from 0.67 – 94.8% of severity on tree. The disease was very severe in Bagalkot with an average severity value of 74.08%. The least average severity (6.73%) was observed in Bellary. Wilt was noticed in a range of 0.1 – 33.3%. The higher percent wilt incidence of 7.59 was observed in Bellary. Anthracnose was recorded in the range of 15-25%, Alternaria 2-15%, Root knot 2-15%, scab 1-5%, fruit rot 2-15%, cercospora was also recorded in the range of 1-2%.

Sharma *et. al.,* (2009) reported present status of pomegranate bacterial blight and its management. All pomegranate cultivars grown in the region viz., Bhagwa, Mirdula, Ganesh, Arkata were susceptible to bacterial blight pathogen. Though the disease exist throughout the year at a temperature range of 9 – 43°C and varying humidities (30- 80%), disease development was rapid during summer rainy months from July to September due to water and high humidity. Mrig bahar crop regulated in June – July (summer rainy season) revealed more disease pressure due to higher infection rate (0.2% unit/day) as compared to Hastabahar crop regulated in September – October (Autumn season) which had infection rate of 0.08/unit/day, there by indicating the brisk spread of the disease in rainy season. Studies on pathogen survivability revealed that *Xanthomonas axonopodis* pv. *Punicae* was able to survive in naturally infection leaves collected from the diseased orchard for one year under laboratory conditions (25.0 – 38.0 °C). Field trial for the management of the bacterial blight reveals that adoption of integrated disease management schedule which includes orchard sanitation, avoiding rainy season crop (Mrig bahar) and regulating hasta bahar crop. Judicious use of antibiotic streptocyclin (500 ppm) in combination with carbendizim (0.15%)/ Mancozeb

(0.2%)/Copper oxy chloride (0.25%)/ at 15 days interval resulted in 82.2% disease control and gave yield of 16 ton/hectare. In separate trial antibiotic Bactronol (2 – Bromo, 2 – Nitropropane, 1-3 diol at 750 – 1000 ppm concentration was also found effective, in managing bacterial blight.

Mulla, *et. al.,* (2009) reported survey and surveillance of bacterial blight (oily spot) disease of pomegranate in Maharashtra, India. Bacterial blight (oily spot) disease caused by *Xanthomonas axonopodis* pv. *Punicae* is one of the worst and baffling disease of pomegranate inflicting qualitative and quantitative losses to the tune of 60-80 percent. Survey was conducted in five district of Maharashtra, Latur, Osmanabad, Ahmednagar, Solapur, Sangli, about 87 orchard surveyed. 70 (80%) orchard affected with bacterial blight/oily spot disease. Highest percentage of disease orchard were found in the district of Sangli (100%) followed by Solapur (78.73%), Latur (73.68%) and Osmanabad (66.66% and in Ahmednagar least (16.6%) orchard were affected.

Subramanayam (2009) reported pomegranate bacterial blight in Andhra Pradesh, district of Anantapur and Mehboobnagar, the highest leaf (71%), stem (81%) and fruit (62.5%) was recorded in Bomanahall mandal and lowest leaf (44.5%) stem (61.6%) and fruit (47.5%) in Gadawal mandal.

Mogle *et. al.,* (2009) reported pathogenecity, morphology and cultural characteristics of *Xanthomonas axonopodis* pv. *Punicae* causing oily spot in pomegranate. Bacterial blight establish on young and healthy plants of highly susceptible variety Ganesh grown in earthen pots in polyhouse. 40 days old thorny leaves of pomegranate inoculated with thorny seeds of sterile *Xanthium strumarium* and nodes by leaf cut method, typical disease symptoms appeared on 13 to 14 days after inoculation symptoms on lower surface are irregular, water soak spots, corresponding to this upper surface of the leaves encircled with yellow halo was observed, later these spot turn brown coalesced and exhibited blight appearance. The re-isolation attempted from artificially infected/disease plant tissue on nutrient agar consistently yielded *X.axonopodis pv. Punicae*. Morphological studies reveal that bacterium was short rod with round ends arranged singly or in pairs, 1.25 to 2 x 0.5 to 0.7 μm in size, motile with single polar flagellum gram negative, capsulated and non-spore former. Cultural characteristics of *Xanthomonas axonopodis* pv. *Punicae* exhibited abundant growth on the culture media viz., Nutrient Agar, Nutrient dextrose agar, Yeast glucose chalk agar, and Oat meal agar. The test bacterium was found to grow better in temperature range of 23 to 30°C and could not grow below 10°C and above 40°C.

Suryawanshi *et. al.,* (2009) studied the efficacy of antibiotic and fungicide against oily spots of pomegranate. The antibiotics viz., streptocyclin and bacterinol and two fungicide carbendizim and copper oxy chloride (@ 250, 400, 500, 1000 ppm) significantly inhibited growth of bacterium over untreated control. At 250, 500 ppm antibiotic streptocyclin, bacterinol were *at par* but caused significant inhibition (16.2 to 22.2%) of the test bacterium over untreated control. However both the antibiotics were found inhibitory at 500 – 1000 ppm. Growth inhibition of 31.1 and 36 per cent was recorded with streptocyclin at 500 ppm and 1000 ppm. This was followed by Bacterinol which recorded 28.3 and 33.6 percent growth. Fungicide copper oxychloride and carbendizim inhibited growth of 25.55 and 24.99 per cent at 500 ppm and 30.55 and 28.32 percent at 1000 ppm. Results in vivo revealed that both the antibiotics evaluated @ 400 ppm and 500 ppm each and fungicide copper oxy chloride (@ 0.25%) and carbendizim (@ 0.1%) recorded significant reduction in disease over untreated control. Antibiotic streptocyclin and Bacterinol recorded significant least mean disease intensity respectively in the range of 5.1 to 7.4 and 5.4 to 7.2 per cent and corresponding mean disease control achieved was 62.6 to 78.1 and 64.6 to 75 percent respectively. Fungicides copper oxy chloride followed by carbendizim were also found effective against disease and recorded mean disease intensity respectively of 6.8 and 8.3 percent and mean disease control of 66.3 and 51.4 percent respectively.

Raghuwanshi *et. al.,* (2009) studied *invitro* evaluation of pesticide/fungicide against *Xanthomonas axonopodis* pv. *Punicae* 29 different formulation consisting of ten fungicides, six antibiotics, two insecticides and one growth hormone were tried. All the treatments have three different concentrations using food poisoning method and using Nutrient sucrose agar (NSA) as basal culture medium. The petriplate, inoculated with bacterium. Results reveal that antibiotic, fungicides, insecticides evaluated at various conc. Inhibited growth of the test bacterium, however one antibiotic, i.e., 2 – bromo 2 nitro propane -1, 3, diol and one fungicide i.e., captan 50 wp were found most effective than the rest of the pesticides. In this two effective chemicals observed under in vitro condition. The chemical which inhibited the growth but mutant colonies formed were copper oxy chloride, cymoxanil + mancozeb, chlorothalonil, metaloxyl + mancozeb, thiram, copper hybroxide, thiomethoxam, mancozeb, potassium metabisulphide, streptocycline, oxytetracyclin, streptopenciline, and the chemical which did not inhibit the growth of the bacterium were fucilazole, dinocap, tridemorphliprobenphos, hexaconozole, difenconazole, propinaconozole, penconazole, carbendizim, benomyl thiophenate methyl, dicofol, imda chloropid, propyalaxin and kasugamycin.

Yenjerapp *et. al.,* (2009) studied the performance of bactericide, bioagents, and botanicals against the bacterial blight of pomegranate. Results shows that lowest disease index of 22.3% in streptocyclin (0.05%) + copper oxychloride (0.2%), followed by bacterinol 100 (0.05%) + copper oxy chloride (0.2%) and bacteriomycin (0.05%) + copper oxychloride (0.2%) which recorded 27.2% disease index. All the three antibiotic exhibited on par efficacy with each other and significantly superior efficacy over *Pseudomonas fluorescence* (0.5%) bioagent and garlic extract (2%) botanical. Highest disease severity was recorded in untreated check plot.

Ravikumar *et. al.,* (2009) studied integrated management of bacterial blight of pomegranate caused by *Xanthomonas axonopodis* pv. *Punicae.* Planting material treated with streptocyclin 500 ppm + copper oxy chloride 2000 ppm eliminated initial inoculum of pathogen. The bioassay studies revealed the effectiveness of antibiotic followed by plant extract. Streptocyclin at 500 ppm was highly effective (20.0 mm inhibition) followed by Bromopal @ 500 ppm (18 mm) and paushamycin @ 500 ppm (18mm) garlic extract @ 5% (15 mm), ocimum @ 5% (12 mm), neem seed extract @ 5% (10 mm) and Onion extract @ 5% (10 mm). In another experiment it revealed that application of streptocyclin @ 500 ppm + copper oxy chloride (2000 ppm) or Bromopol @ 500 ppm + copper oxychloride (2000 ppm) sprayed eight times at an interval of 10 days followed by zinc sulphate + magnesium sulphate (0.1%) + boron (0.1%) + lime sulphate (0.1%) were found effective in integrated management. The streptocyclin 500 ppm + copper oxychloride 2000 ppm and Bromopol 500 ppm + copper oxychloride 2000 ppm were found effective in controlling the disease by recording 6.92 and 18.7 percent incidence respectively when compared to control 95.3% in 2006 and 8.2 and 22.1 percent and 92.5% respectively in 2007 after the 8th spray. The maximum yield of 8.8 tones/ha and 8.49 tones/ha was recorded in streptocyclin 500 ppm followed by Bromopal 500 ppm + copper oxychloride 2000 ppm in 2006 and 7.45 tones/ha and 7.85 tones/ha. The untreated check recorded yield of 0.5 and 1.8 tones/ha during 2006 and 2007 respectively. The integrated disease management strategy included field sanitation, disinfecting the secature with 1.5% sodium hypochloride and smearing the stem with streptocycline @ 0.5 g and copper oxychloride 3.0 g along with red oxide @ 200 g per liter of water, dusting with bleaching powder @ 25 kg/ha soil surface around the pland and Sept – Oct pruning was found effective to bring down the initial inoculum of pathogen in the garden.

Gopalkrishna and Saxena, (2009) revealed that under the Mission of NHM Network project on "Mitigating the bacterial blight disease of pomegranate in Maharashtra, Karnataka and Andhra Pradesh", demonstration in the technology for the management of bacterial blight disease of pomegranate were taken up at three locations of the Southern Karnataka. The crop was taken up during Sept – Oct, 2009 (Hasta Bahar). The recommended cultural operations, need based plant protection application right from pruning were implemented. Weekly observations were recorded on severity of bacterial blight disease of leaves, stems and fruits, before and after the crop was subjected to the technology package. Simultaneously, periodical survey and surveillance of the bacterial blight disease was also undertaken in other orchards under pomegranate cultivation. Weather data in respective locality were also recorded. It was observed that before implementing the technology, the per cent bacterial blight incidence (PPI) was very severe in all the three plots and varied from 53.4 in Jagalur to 69.5 in Sira, where as in Hiriyur the PPI was 1.50, thus indicating that the technology was very effective mitigating the severity of the disease.

Somasekhara and Gaddanakeri, (2009), reported that, severe incidence of pomegranate wilt due to *Ceratocystis fimbriata* rendered vast stretches of orchard soil sick. Farmers uprooted severely wilted pomegranate trees and gap filled with various other hosts. The plants thus planted were *Achras sapota, Carica papaya, Citrus aurantifolia, Curcuma longa, Ficus elastica, Mangifera indica, Murrayya koenigil, Piper betle, Psidum guajava, Saccharaum officinarum, Tectona grandis, Vitis vinifera, Zizyphus mouritiana* and *Punica granatum.* The fungus *C. fimbriata* was artificially incolulated to the plants through air layering. Wilt incidence, if any was observed during August 2003 (after 27 months). There was no infection of *C. fimbriata* in the inoculated stem portion and also there was no root infection in the plants. The check pomegranate plants Cv. Mridula, Kesar, G-137 expressed blue stain in the inoculated location and also plants wilted within six months. These results showed that, the fungus *C. fimbriata* infects pomegranate only.

Somashekara (2009) studied in vitro evaluation of fungicides for the management of wilt pathogen (*Ceratocystis fimbriata*). Pomegranate is affected with many diseases among them wilt *Ceratocystis fimbriata* is an important problems the severity was reported up to 5.31%. An attempt was made to evaluate fungicides against this pathogen in vitro and in vivo by adopting various techniques. The various fungicides were selected and mixed with potato dextrose agar medium with different concentration and placed fungal disc or infected bits on the medium. The fungal growth survived of *C. fimbriata* on PDA medium was completely inhibited in fungicide Bavistin, Benomyl, Mancozeb, Ridomil, Topsin, Ziram and Tilt mixed medium at 1000 ppm concentration over control (23.34 mm). In another experiment powder of Black soil was mixed with fungicide at different concentration (400, 600, 800, 1000 ppm) in 100 g of soil. The fungust disc (2mm) or infected bits (2mm) were placed at the centre of the fungicide mixed soil. The fungus disc or infected bits were removed from the soil after 30 days of incubation and washed with sterilized distilled water, washed disc or infected bits were placed on PDA medium and recorded survival ability of *C. fimbriata.* The fungicides viz., Bayleton, Mancozeb, Ridomil, Ziram Tilt completely inhibited the growth of *C. fimbriata* on PDA medium. The fungicides Bavistin, Benomyl and Thiophenate methyl heated soil did not inhibited the growth of the fungus. The survival ability of the fungus in control was 92.13%. The fungal dry matter was studied in potato dextrose broth mixed with various fungicides at 600-1000 ppm. After 15 days of incubation the fungus was collected on Watsman No.44 filter paper and oven dried. The least fungus dry matter was obtained in the fungicide Benomyl (50 mg) Calixin (40 mg) Fostyl (40 mg), Tilt (0.0 mg) and Ziram (30 mg) mixed solution at 1000 ppm concentration and dry matter weight in untreated fungicide solution was 170 mg. Based on the above experiment, the effective fungicide were tried in field condition at various locations from 1999 to 2005 to re-establish the infected gardens. The infected plants were removed and the soil was sterilized with Formalin solution before new planting and the effective fungicide propiconazol (0.1%) along with Boric acid (25 gm/plant) was applied to partially infected and surrounding plants. Totally 338 plants were recovered out of 345 re planted plants and there was no further spread of disease in the field. The soil sterilization with Formalin (4%) in the infested area along with garden was very useful to protect the plants from the disease.

Navale *et. al.,* (2009) studied the chemical control of fungal leaf and spot disease of pomegranate Cv. Mirdula. *Alternaria alternate, Colletotrichum gloeosporioides, Cereospora granatum, Drechslera rostrata* are the major fungal leaf and fruit spots causing organism under agroclimatic conditions of western Maharashtra, India. The treatment included fungicide carbendizim @ 0.1%, propiconizol @ 0.1%, copper oxy chloride @ 0.4%, Thiophenate Methyl @ 0.1%, Mancozeb @ 0.3%, Difenconazol @ 0.1%, Penaconazol @ 0.1% were selected for control of leaf spot diseases of pomegranate. The first prophylactic spray of selected fungicide was given to young leaves after pruning and subsequent need based (9 sprays) were given at an interval of 10 days. The pooled data indicated that the fungicide difenconazol @ 0.1% showed significant least per cent disease intensity on leaves (1.08%) and fruits (6.93%) and maximum percent disease control on leaves (89.79%) and fruits (91.07%) followed by propicanozol @ 0.1% (1.61 and 8.58%) PDI on leaves and fruits respectively) and carbendizim @ 0.1% (1.83 and 9.35%) PDI on leaves and fruits respectively). The highest fruit yield was obtained with difenconazole spray. The results of the residue analysis indicated that the residues of difenconazole, propiconazole and carbendizim were below detectable level in the arils at 1st, 3rd and 5th day after spray.

Kishore Khosla *et. al.,* (2009) studied the management of leaf and fruit disease of pomegranate through chemicals. The intensive cultivation of pomegranate a number of fungal have been found associated causing fruit spot and rot diseases in many varieties of pomegranate. The fungal pathogen associated with fruit spot diseases are *Drechslera rostrata, Baltraneilla humicola and wilt fruit rots are Glomeralla cingulata, (Colletotrichum gloesporioides, Phomopsis aucubicoal, Cladosporium oxysporium, Aspergillus sps, Rhizopus sps,* in addition to this bacterial blight caused by *Xanthomonas axonopodis pv. Punicae* has also been recorded mainly in soft seeded variety Mirdula, Ganesh, Sindhuri (Bhagwa), Arakta. The infected leaves shed prematurely and subsequent emerging leaves also get infested. The bacterial infection on leaves and fungal infection on fruits considerably reduce the marketable yield causing huge losses to the growers. For the management of fruit spot and rot diseases, six different spray schedules were tried viz., schedule VI (Companion – copper oxy chloride – captan – companion), II (captan – companion – copper oxychloride – companion) and V (carbendizim – mancozeb – captan – companion) were highly effective and provided more than 80 percent control of spot and rot diseases. Maximum yield (16.8 tonne/ha) was recorded in trees covered under schedule VI followed by 16.4 tonne/ha under schedule II and 15.3 tonne/ha in schedule V as compared to control 2.6 tonne/ha. For the management of bacterial blight a combination of copper oxy chloride (0.2%) and streptocyclin 0.01% proved most effective in comparison to individual application of either copper oxy chloride (0.3%) or streptocyclin (0.01%) when sprayed five times at an interval of 12 days.

Prashanth and Satareddi (2009) studied the variability in different isolates of anthracnose of pomegranate caused by *Colletotrichum gloeosporioides* has become potentially destructive disease both under field and post harvest storage conditions. In recent years the disease has become a serious problem in northern Karnataka various aspects of the pathogen with respect to cultural and pathogenic variation in order to understand the pathogen variability for its effective management of this menace. All the eight isolates of *C. gloesporioides* collected from Northern Karnataka (India) showed significant difference in relation to radial growth, colony character, pigmentation and sporulation on potato dextrose agar. Maximum radial growth of 90 mm and 88.5 mm was recorded in Cgr-2 and Cgr-1 isolates respectively. It was followed by Cgb-2 and Cgk-1 which recorded 87.00 mm and 85.00 mm radial growth, with respect to colony character three types of colony were observed, maximum sporulation with excellent grades was recorded in Cgb-1 while fair sporulation was noticed in Cgk-1, Cgk-2 and Cgk-3 isolates. The size of spores also varied among the isolates. Large sized spores (2.25 – 2.50 x 0.90 – 1.0 μm) were observed in Cgk-1, while least size spores (1.37 – 1.5 x 0.87 – 1.0 μm) were recorded in Cgb-1 isolates.

Mallesh *et. al.,* (2009) studied native plant growth promoting rhizobacteria for the suppression of pathogen of pomegranate (*Punica granatum L*). *Ceratocystis fimbriata, Colletotrichum gleosporioides and Xanthomonas axonopodis pv. Punicae* are devastating pathogen threatening pomegranate cultivation in South India. The farmer causes vascular wilt and later two form lesions on foliage, stem, twigs and fruits leading to general deliberation of the tree. They also produced fruit cracking symptoms leading to severe reduction in market value of the fruit. They investigated the ability of seven selected native rhizobacterial strains (designated RB1, RB10, RB13, RB31, RB43, RB50) isolated from healthy rhizospheres and Rhizopheres (of a couple of Horticultural crops) of the region to suppress the activity in vitro of fore mentioned pathogen using the dual cultural technique percent myeclial inhibition due to these seven rhizobacteria was computed in respect of *C. fimbriata* and *C. gloeosporioides* and zone of inhibition produced by threes rhizobacteria against *X. axonopodis pv. Punicae* was measured. RB 50 was the most effacious of all as it produced a maximum zone of inhibition (19mm) against the bacterial pathogen and recorded highest per cent mycelial inhibition against the fungal pathogen (67.7% - 72.2%) against *C.fiambriata* and *C. gloeosporioides* respectively. RB50, RB31, RB13 (In that order) appreciably suppressed the growth of both *Xanthomonas axonopodis pv. Punicae* and *C. gloeosporioides* while RB50, RB43, RB1 registered > 60 percent mycelial inhibition of *C. fimbriata*. These PGPR strains their plant growth promoting ability have been earlier tested by authors hold a lot of promise either alone or in a consortium towards combating these diseases of pomegranate. Phenotypic characterization showed that these seven potent PGPR isolates to be fluorescent pseudomonads.

SataReddy *et. al.,* (2009) studied the role of bio agents and botanicals in the management of anthracnose of pomegranate. Pomegranate is having serious diseases like *Colletotrichum gleosporioides*, in the present study PGPR efficacy bio agents and botanicals in the management of the disease which were studied invitro and invivo. There was significant difference among bio agents tested. *Trichoderma viride* (1) (79.1%) was found to be significant in inhibiting the mycelial growth of *C. gloeosporioides* followed by *Triochoderma herzianum* (1) 62.8%. Among the bacterial bio agents *Pseudomonades fluorescence* recorded maximum inhibition (54.8%) and least inhibition of mycelial growth of C. *gloeosporioides* was recorded in *Bacillus subtilis* (34.9%). Maximum mycelium growth inhibition was noticed (93.1 and 88.1%) in Eucalyptus extract and Garlic extract at 10% concentration, respectively. Effectiveness of botanical & bio-agents which performed well under laboratory conditions, were tested in field condition to manage the anthracnose of pomegranate. Among the seven treatments minimum disease index was recorded in *T. viride* (18.9%) treatment with highest yield (7290 kg/ha).

Somasekara *et. al.,* (2009) studied the biological management of root knot nematode (*Meloidogyne incognita*) infesting pomegranate in Karnataka. Root knot nematode was first reported in 1922 in Spain. The predominant root knot nematodes which affecting pomegranate is *M. incognita, M. acrita, M. javanica*. Heavy root galling and visible damage to pomegranate trees in young orchards under irrigations are frequently encountered *M.incognita* commonly occurs in Karnataka, Andhra Pradesh, Maharashtra, North Gujarat. This root know nematode was noticed on both wilting agents (*Ceratocystis fimbriata)* and also non-wilted plants. Root knot nematode infestation may be helpful for the entry of wilt pathogen (*Ceratocystis fimbriata L*) in to root system. The bio agents *Paecilomyces lilacinus* (20 g/plant) and *Verticillium chlamydosporium* (20 g/plant) along with carbofuran (20 g/plant). *Paecilomyces lilacinus* recorded lowest nematode population, also root gall number both at 45 and 180 days of sampling. The application of carbofuran also equally reduced population of *Meloidogyne incognita* and root galls with increase in yield. The soil application of bio-agents *P.lilacinus* or nematicide carbofuran at the rate of 20 g/plant is very useful in reducing root knot nematode population and also root gall number in infested pomegranate plants and may protect the plant from entry of wilt pathogen, *Ceratocystis fimbriata*.

References

Abdol Karim Zarei, Zabiholah Zamani, Mohammad Reza Fatahi Moghadam, 2009, SSR Analysis of progenies derived from pollination of pomegranate Cv. Malase – Torshe – Saveh, 2nd International Symposium on "Pomegranate and Minor including Mediterranean Fruits" from June 23-27, 2009 held at UAS, Dharwad, Karnataka, India, pp.71.

Alburatamani, S. K.S. and Fatope, M. O., 2006, Nutritive potentials of *Punica granatum* seed soil, Ist International Symposium on "Pomegranate and Minor fruits Mediterranean Fruits" from October 16-19, 2006 held at Cukurova University, Adana, Turkey.

Ali Mohseni, 2009, Introducing the Disease of Aril De-colouring of Pomegranate and Green Drying of the Pomegranate Tree and Difficulties created in Pomegranate Orchards in Iran, 2nd International Symposium on "Pomegranate and Minor including Mediterranean Fruits" from June 23-27, 2009 held at UAS, Dharwad, Karnataka, India, pp.144.

Amerine, M.A. Berg, I.I.W. and Cruness, W.V., 1972, The Technology of Wine Making 3rd edition, AVI Publishing Company, West Port, Connecticut, PP 126-132

Amos Levin., 2009, Practical Applications of Cold Storage of Pomegranates in a Controlled Atmosphere in Israel, 2nd International Symposium on "Pomegranate and Minor including Mediterranean Fruits" from June 23-27, 2009 held at UAS, Dharwad, Karnataka, India, pp.122.

Anitha Kumari, Lakshmi, B.K.M., SureshKumar, T. and Satyanarayana Reddy, G., 2009, Population Dynamics of Major pests of Pomegranate in Central Telangana Zone of Andhra Pradesh, 2nd International Symposium on "Pomegranate and Minor including Mediterranean Fruits" from June 23-27, 2009 held at UAS, Dharwad, Karnataka, India, pp.148.

Anonymous., 2006, Fruit juice consumption habits of turkey and the world and Pomegranate juice, 2nd International Symposium on "Pomegranate and Minor Mediterranean Fruits" from October 16-19, 2006 held at Cukurova University, Adana, Turkey.pp.125.

Aradhya, M., Stover, E., Koehmstedt, A. and Moersfelder, J., 2006, Low genetic diversity and lack of genetic differentiation in Punica as evidenced by amplified fragment length polymorphisms, Ist International Symposium on "Pomegranate and Minor Mediterranean Fruits" from October 16-19, 2006 held at Cukurova University, Adana, Turkiye, pp.64.

Ataseven Isik, E., 2006, Similarities between pomegranate parents and crosses regarding some fruits characteristics, I[st] International Symposium on "Pomegranate and Minor Mediterranean Fruits" from October 16-19, 2006 held at Cukurova University, Adana, Turkey.pp.67.

Atta Hosseini, S.A., Rezaian, S. and Zabihi, H.R., 2006, Determination of fruit set percentage from pomegranate flowers in the Shisheh – cap variety of Ferdows region in Iran, I[st] International Symposium on "Pomegranate and Minor Mediterranean Fruits" from October 16-19, 2006 held at Cukurova University, Adana, Turkey.pp.51.

Aytelun polat, A. and Caliskan, O., 2006, Effect of indolebutyric acid (IBA) on rooting of cutting in various pomegranate varieties, I[st] International Symposium on "Pomegranate and Minor Mediterranean Fruits" from October 16-19, 2006 held at Cukurova University, Adana, Turkey.pp.50.

Bagle, B.G., 2009, Studies on Varietal Reaction, Extent of Damage and Management of Anar Butterfly *Virachola isocrates* in Pomegranate, 2nd International Symposium on "Pomegranate and Minor including Mediterranean Fruits" from June 23-27, 2009 held at UAS, Dharwad, Karnataka, India, pp.145.

Balikai, R.A., Prasanna, P.M. and Kotikal, Y.K., 2009, Status of pomegranate pests and their management strategies in India, 2nd International Symposium on "Pomegranate and Minor including Mediterranean Fruits" from June 23-27, 2009 held at UAS, Dharwad, Karnataka, India, pp.147.

Baptist, B.A. 1944. The Fruit-Piercing Moth (*Othreis fullonica* L.) with Special Reference to its Economic Importance. Indian J. Entomology. 6: 1-13.

Bharati, N., 2006, Micro-propagation of pomegranate in large scale, I[st] International Symposium on "Pomegranate and Minor Mediterranean Fruits" October 16-19, 2006 held at Cukurova University, Adana, Turkey.pp.73.

Bar-Ya'akov,, I., Trainin, T., Hefetz, H., Hatib, K. and Holland, D., 2006, Improving pomegranate cultivars in Israel, I[st] International Symposium on "Pomegranate and Minor Mediterranean Fruits" from October 16-19, 2006 held at Cukurova University, Adana, Turkey.pp.2.

Bayram, E. and Dundar, V., 2006, Effect of Different Storage conditions on quality of Pomegranate Fruits (cv. Hicaznar), I[st] International Symposium on "Pomegranate and Minor Mediterranean Fruits" from October 16-19, 2006 held at Cukurova University, Adana, Turkey.pp.114.

Benagi, V. I., Ravikumar, M. R., Gowda, S. B. and Basavaraj, B.B., 2009, Survey on Diseases of Pomegranate in Northern Karnataka, 2nd International Symposium on "Pomegranate and Minor including Mediterranean Fruits" from June 23-27, 2009 held at UAS, Dharwad, Karnataka, India, pp.135.

Bhosale, V. P., Jadav, R.G., Masu, M. M., Sitapara, H. H. and Patel, H.C., 2009, Response to Different Media and Plant Growth Regulators on Rooting and Survival of Airlayers in Pomegranate (*Punica granatum L*) Cv. Sindhuri, 2nd International Symposium on "Pomegranate and Minor including Mediterranean Fruits" from June 23-27, 2009 held at UAS, Dharwad, Karnataka, India, pp.73.

Bignami, C., Cammilli, C., Ciaccia, C., Cristofore, V. and Roviglioni, R., 2006, Collection conservation and valorization traditional fruit trees varieties for niche market development : the case of pomegranate in Southern Lazio (Italy), I[st] International Symposium on "Pomegranate and Minor Mediterranean Fruits" from October 16-19, 2006 held at Cukurova University, Adana, Turkey.pp.121.

Bikramjit Singh, Sukdev Singh and Gurjaipal Singh., 2009, Influence of Planting Time and IBA on Rooting and Growth of Pomegranate (*Punica granatum L*). 2nd International Symposium on "Pomegranate and Minor including Mediterranean Fruits" from June 23-27, 2009 held at UAS, Dharwad, Karnataka, India, pp.73.

Bosch, J.E. 1971a. Fruit Piercing Moth Research. Rhodesia Agric. J. 68: 19-21.

Bosch, J.E. 1971b. The Possibility of Controlling Fruit Piercing Moths by Means of an Odor Repellent. Rhodesia Agric. J. 68: 113.

BŠnziger, H. 1982. Fruit-Piercing Moths (Lep., Noctuidae) in Thailand: A General Survey and Some New Perspectives. Mitteilungen der Schweizerischen Entomologischen Gesellschaft. 55: 213-240.

Chaurasia, A. K., Subramaniam, V.R., Balkrishna and Sane, P. V., 2009, RAPD Based Genetic Diversity among Indian and Exotic varieties of pomegranate, 2nd International Symposium on "Pomegranate and Minor including Mediterranean Fruits" from June 23-27, 2009 held at UAS, Dharwad, Karnataka, India, pp.71.

Cinar, J. and Inanc, L., 2006, Storage stability of seed powder and black berry added pomegranate juice, I[st] International Symposium on "Pomegranate and Minor Mediterranean Fruits" from October 16-19, 2006 held at Cukurova University, Adana, Turkey.pp.107.

Comstock, J.A. 1963. A Fruit-Piercing Moth of Samoa and the South Pacific Islands. Canadian Entomologist. 95(2): 218-222.

D'Aquino, S., Schirra, M., Palma, A., Angioni, A. Cabras, P., Gentile, A. and Tribulato, E., 2006, Effectiveness of Fludioxonil in control storage Decay on Pomegranate fruit, I[st] International Symposium on "Pomegranate and Minor Mediterranean Fruits" from October 16-19, 2006 held at Cukurova University, Adana, Turkey.pp.112.

Dan Rymon., 2009, Mapping Features of the Global Pomegranate Market, 2nd International Symposium on "Pomegranate and Minor including Mediterranean Fruits" from June 23-27, 2009 held at UAS, Dharwad, Karnataka, India, pp.153.

Dheeraj Singh, Wangchu, L., Chaudhary, M., Dayal, H. and Meena, M. L., 2009, Drying of Pomegranate Seeds (Anardana) under different conditions, 2nd International Symposium on "Pomegranate and Minor including Mediterranean Fruits" from June 23-27, 2009 held at UAS, Dharwad, Karnataka, India, pp.125.

Dhemre, J. K., Desale, S. B., Kokate, K. D. and Ahire, M. C., 2009, A Study on Knowledge of Progressive Farmers on Pomegranate Production Technology, 2nd International Symposium on "Pomegranate and Minor including Mediterranean Fruits" from June 23-27, 2009 held at UAS, Dharwad, Karnataka, India, pp.154.

Dhemre, J. K., Desale, S. B., Kokate, K. D. and Ahire, M.C., 2009, Impact of KVK Training Programme on Knowledge and Adoption of Bahar Treatment in Pomegranate in Tribal

Belts of Maharashtra, 2nd International Symposium on "Pomegranate and Minor including Mediterranean Fruits" from June 23-27, 2009 held at UAS, Dharwad, Karnataka, India, pp.155.

Dhillon, W.S., Gill, P.P.S. and Navprem Singh., 2009, Effect of Nitrogen, Phosphorus and Potassium Fertilisation on Growth, Yield and Quality of Pomegranate Cv. Kandhari, 2nd International Symposium on "Pomegranate and Minor including Mediterranean Fruits" from June 23-27, 2009 held at UAS, Dharwad, Karnataka, India, pp.106.

Edward, M. Raja, 2008, Identification of important micronutrient disorders in Pomegranate in Maharashtra, Dalimb Virut Samarnika, pp 149-151.

Ekbote, S. D., 2008, Occurrence of Root Knot Nematode on Pomegranate in Karnataka, Dalimb Virut Samarnika, pp 140.

Ephraim Lansky, M.D., 2006, A premier research and development company for Pomegranate, Ist International Symposium on "Pomegranate and Minor Mediterranean Fruits" from October 16-19, 2006 held at Cukurova University, Adana, Turkey.pp.80.

Gaikwad, R.S. and Dhemre, J.K., 2009, Studies on preparation of Anardana from Pomegranate Arils, 2nd International Symposium on "Pomegranate and Minor including Mediterranean Fruits" from June 23-27, 2009 held at UAS, Dharwad, Karnataka, India, pp.126.

Ganji Moghadam, E. and Nikkhah, Sh., 2006, Effect of post harvest heat treatments in reducing the pomegranate moth damage, Ist International Symposium on "Pomegranate and Minor Mediterranean Fruits" October 16-19, 2006 held at Cukurova University, Adana, Turkey.pp.110.

Gawade, B.J., More, H.G., Kad, V.P. and Konwade, V.L., 2009, Effect of Intermittent Drying Techniques on Quality of Anardana, 2nd International Symposium on "Pomegranate and Minor including Mediterranean Fruits" from June 23-27, 2009 held at UAS, Dharwad, Karnataka, India, pp.122.

Gazel, M., Caglayon, K., Ulubas Serce, C. and Durgac, C., 2006, Detection of hop stunt viroid and cucumber mosaic virus in Pomegranate (*Punica granatum L.*) trees in the East Mediterranean region of Turkiye, Ist International Symposium on "Pomegranate and Minor Mediterranean Fruits" October 16-19, 2006 held at Cukurova University, Adana, Turkey.pp.93.

Gentile, A., Domina, F., Finocchiaru, E., Lamalya, S. and Tribulato, 2006, Pomegranate germplasm evaluation in Sicily, Ist International Symposium on "Pomegranate and Minor Mediterranean Fruits" from October 16-19, 2006 held at Cukurova University, Adana, Turkey.pp.27.

Ghasemnegad, M., Ranjbar, H. and Sarkhosh, A., 2006, Influence of Methyle Jasmonate on inducing chilling tolerance in Pomegranate fruit (Malas Save), Ist International Symposium on "Pomegranate and Minor Mediterranean Fruits" from October 16-19, 2006 held at Cukurova University, Adana, Turkey.pp.119.

Gill, P. P. S, Dhillon, W. S. and Navprem Singh., 2009, Influence of Training Systems on Growth, Yield and Fruit Quality of Pomegranate cv 'Kandhari', 2nd International Symposium on "Pomegranate and Minor including Mediterranean Fruits" from June

23-27, 2009 held at UAS, Dharwad, Karnataka, India, pp.101.

Goel, M.C., Singh, K. J. B and Bhende, S.N., 2009, Response of Application of customized Fertilizer Grade (CFG) on Yield and Quality of Pomegranate, 2nd International Symposium on "Pomegranate and Minor including Mediterranean Fruits" from June 23-27, 2009 held at UAS, Dharwad, Karnataka, India, pp.107.

Gopala Krishna. C. and Saxena, A.K., 2009, Management of Bacterial Blight Disease of Pomegranate in Southern Districts of Karnataka, 2nd International Symposium on "Pomegranate and Minor including Mediterranean Fruits" from June 23-27, 2009 held at UAS, Dharwad, Karnataka, India, pp.139.

Hadj Salem, H., Hasnaovi, N. Brini, W. and Mart, M., 2006, Quality assessment and use of juice from local acid Pomegranate (*Punica granatum L*) cultivars in Tunisia, Ist International Symposium on "Pomegranate and Minor Mediterranean Fruits" from October 16-19, 2006 held at Cukurova University, Adana, Turkey.pp.124.

Hargreaves, E. 1936. Fruit-Piercing Lepidoptera in Sierra Leone. Bull. Entomol. Res. 27: 589-605.

Hasnaoui, N., Mars, M., Chibani, J. B., Marrakchi, M. and Trifi, M., 2006, Molecular polymorphism in Tunisian pomegranate (*Punica granatum L*) germplasm as revealed by RAPD Fingerprints, Ist International Symposium on "Pomegranate and Minor Mediterranean Fruits" from October 16-19, 2006 held at Cukurova University, Adana, Turkey.pp.70.

Hasnaoui, N., Sebastiani, F., Buonamici, A., Mars, M., Trifi, M., Marrakchi, M. and Vendramin, G. G., 2006, Construction of an enriched genomic library of pomegranate (*Punica granatum L*) and design of SSR primers, Ist International Symposium on "Pomegranate and Minor Mediterranean Fruits" October 16-19, 2006 held at Cukurova University, Adana, Turkey.pp.69.

Hasnaovi, N., Mars, M., Trifi, M., Melgarejo Moreno, P., Marrakchi, M. and Hernandez Grecia, F., 2006, Juice chemical composition of pomegranate cultivars (*Punica granatum L.*) grown in Tunisia, Ist International Symposium on "Pomegranate and Minor Mediterranean Fruits" from October 16-19, 2006 held at Cukurova University, Adana, Turkey.pp.135.

Hepaksoy, S., Engin, H., Kukul, Y. S. and Erogul, D., 2006, The effect of irrigation levels on physiological response in some pomegranate (*Punica granatum L*) varieties, Ist International Symposium on "Pomegranate and Minor Mediterranean Fruits" from October 16-19, 2006 held at Cukurova University, Adana, Turkey.pp.53.

Hepaksoy, S., Sen, F., Erogul, D. and Aksoy, U., 2006, Antioxidant capacity of some Pomegranate (*Punica granatum L.*) varieties, Ist International Symposium on "Pomegranate and Minor Mediterranean Fruits" October 16-19, 2006 held at Cukurova University, Adana, Turkey.pp.75.

Heu, R., K. Teramoto, E. Shiroma, B. Kumashiro, and R. Dinker. 1985. Hawaii Pest Report. 5(1-4): 4-5.

Hiwale, S. S., 2008, Rejuvenation of Pomegranate orchards through non-selective pruning, Dalimb Virut Samarnika, pp 121-122.

Hiwale, S. S., More, T. A. and Bagle, B.G., 2009, Root Distribution pattern in Pomegranate Cv. Ganesh, 2nd International Symposium on "Pomegranate and Minor including Mediterranean Fruits" from June 23-27, 2009 held at UAS, Dharwad, Karnataka, India, pp.106.

Hosamani, S. B., Mamle Desai, N. R. and Kulkarni, B. S., 2008, Promotion of Farmer Market Linkages in Pomegranate, Dalimb Virut Samarnika, pp 162-164.

Hosamani, S. B., Hiremath, G. M. and Kulkarni, B. S., 2009, Indian Pomegranate Export in the pre and post WTO Regime, 2nd International Symposium on "Pomegranate and Minor including Mediterranean Fruits" from June 23-27, 2009 held at UAS, Dharwad, Karnataka, India, pp.153.

Irit Bar-Ya'akov, Kamel Hatib, Talia Nadler-Hassar, Zohar Ben-Simhon, Taly Trainin, Hamutal Borochov Neori and Doron Holland, 2009, Breeding for improved cultivars using molecular knowledge on diversity in Pomegranate, 2nd International Symposium on "Pomegranate and Minor including Mediterranean Fruits" from June 23-27, 2009 held at UAS, Dharwad, Karnataka, India, pp.70.

Islam, A., Yarilgac, T. and Ozguven, A. I., 2006, Pomological and morphological characteristics of pomegranate grown in Black sea region, Ist International Symposium on "Pomegranate and Minor Mediterranean Fruits" from October 16-19, 2006 held at Cukurova University, Adana, Turkey.pp.29.

Jadhav, M. S., Mali, M. D. and Waliwal, V. D., 2008, Economics of production and Marketing of Pomegranate in Satara District of Maharashtra, Dalimb Virut Samarnika, pp 155-161.

Jaganath, S., Meenakshi, H.C., Harinikumar, K.M. and Nachegowda, V., 2009, Effect of Microbial Inoculants on Rooting of Pomegranate (*Punica granatum L*) cvs. Bhagwa and Ganesh stem cuttings, 2nd International Symposium on "Pomegranate and Minor including Mediterranean Fruits" from June 23-27, 2009 held at UAS, Dharwad, Karnataka, India, pp.72.

Jaganath, S., Najibullah Mujadadi and Nachegowda, V., 2009, Regeneration of pomegranate (*Punica granatum L*) Ruby Mridula stem cutting with Microbial Inoculants, 2nd International Symposium on "Pomegranate and Minor including Mediterranean Fruits" from June 23-27, 2009 held at UAS, Dharwad, Karnataka, India, pp.72.

Jahagirdar, S. Jamadar, M. M., Patil, D.R. and Ranganath, M. C., 2006, Exploitation of new molecules in management of fruit spot of pomegranate incited by *Colletotrichum gleosporiorides* in India, Ist International Symposium on "Pomegranate and Minor Mediterranean Fruits" from October 16-19, 2006 held at Cukurova University, Adana, Turkey.pp.97.

Jamadar, M. M., Jawadagi, R. S., Sataraddi, A. R., Patil, D. R. and Patil, R. V., 2009, Status of Pomegranate Disease and their Management in Northern Karnataka in India, 2nd International Symposium on "Pomegranate and Minor including Mediterranean Fruits" from June 23-27, 2009 held at UAS, Dharwad, Karnataka, India, pp.134.

Jbir, R., Hasnaoui, N., Trifi, M., Marrakchi, M. and Mars, M., 2006, Molecular characterisation in Tunisian Pomegranate (*Punica granatum L*) cultivars as revealed by AFCP, I[st] International Symposium on "Pomegranate and Minor Mediterranean Fruits" from October 16-19, 2006 held at Cukurova University, Adana, Turkey.pp.71.

Joshi, V.R., Masalkar, S. D., Kulkarni, S.R. and Patil, S.D., 2009, Effect of different chemicals on leaf shedding of Pomegranate, 2nd International Symposium on "Pomegranate and Minor including Mediterranean Fruits" from June 23-27, 2009 held at UAS, Dharwad, Karnataka, India, pp.110.

Kad, V.P., More, H.G., Gawade, B. J. and Kanawade, V.L., 2009, Studies on Extraction, Clarification, Preservation and Storage of Pomegranate (*Punica granatum L*) juice, 2nd International Symposium on "Pomegranate and Minor including Mediterranean Fruits" from June 23-27, 2009 held at UAS, Dharwad, Karnataka, India, pp.128.

Kevin R. Day and Erik D. Wilkins, 2009, Commercial Pomegranate (*Punica granatum L*) production in California, 2009, 2nd International Symposium on "Pomegranate and Minor including Mediterranean Fruits" from June 23-27, 2009 held at UAS, Dharwad, Karnataka, India, pp.99.

Kezban Yazici, Salih Ulger and Lomi Kaynak, 2009, The Relationship between Flower Development and Endogenous Gibberellins in Pomegranate (*Punica granatum L*) Cv.Hicaznar, 2nd International Symposium on "Pomegranate and Minor including Mediterranean Fruits" from June 23-27, 2009 held at UAS, Dharwad, Karnataka, India, pp.109.

Khajuria, D.R., 2009, Outbreak of pomegranate mite, *Tenuipalpus granati* sayed on pomegranate in Kullu District of Himachal Pradesh, 2nd International Symposium on "Pomegranate and Minor including Mediterranean Fruits" from June 23-27, 2009 held at UAS, Dharwad, Karnataka, India, pp.151.

Khorsandi, F. and Yazdi, F.A., 2006, Improvement of pomegranate fruit quality and phytoestrogen content by zinc fertilization, I[st] International Symposium on "Pomegranate and Minor Mediterranean Fruits" from October 16-19, 2006 held at Cukurova University, Adana, Turkey pp.76.

Khoshnam, F., Tabatabaeefar, A. and Bordhei, A., 2006, Mass prediction of pomegranates based on its dimension, I[st] International Symposium on "Pomegranate and Minor Mediterranean Fruits" from October 16-19, 2006 held at Cukurova University, Adana, Turkey pp.137.

Khoshroo, A., Zoroofi, R.A., Keyhani, A., Zamani, Z. and Rafiee, Rh., 2006, Pomegranate quality evaluation using machine vision, I[st] International Symposium on "Pomegranate and Minor Mediterranean Fruits" October 16-19, 2006 held at Cukurova University, Adana, Turkey pp.123.

Kishore Khosla, Jayant Kumar, K.S., Bharadwaj. S.S, 2009, Chemical control of Fungal Leaf and Fruit spot Diseases of Pomegranate, 2nd International Symposium on "Pomegranate and Minor including Mediterranean Fruits" from June 23-27, 2009 held at UAS, Dharwad, Karnataka, India, pp.141.

Koshi – Dehkordi, M., Sayed – Tabatabaei, B.E. and Bahar, M., 2006, Characterization of microsatellites and development of SSR marker in pomegranate, I[st] International Symposium on "Pomegranate and Minor Mediterranean Fruits" from October 16-19, 2006 held at Cukurova University, Adana, Turkey pp.66.

Kotikal, Y.K., Ananda, N. and Balikai, R.A., 2009, Seasonal Incidence of Major Sucking Pests of Pomegranate and their Relation with weather parameters in India, 2nd International Symposium on "Pomegranate and Minor including Mediterranean Fruits" from June 23-27, 2009 held at UAS, Dharwad, Karnataka, India, pp.151.

Krishna Naik., L., Jagginavar, S. B. and Biradar, A. P., 2009, Beetle Enemies of Pomegranate and their Management, 2nd International Symposium on "Pomegranate and Minor including Mediterranean Fruits" from June 23-27, 2009 held at UAS, Dharwad, Karnataka, India, pp.146.

Krishna chandra and Greep, S., 2005, Biocontrol agents and Biopesticides, Published by Regional Centre of Organic Farming, Bangalore.

Kucuk, E., Tanrisever, A., Safak, C., Ozkarakas, I., Gurnil, K. and Tokmak, S., 2006, Fruit set and self fruitfulness of some pomegranate (*Punica granatum L*) cultivars, I[st] International Symposium on "Pomegranate and Minor Mediterranean Fruits" from October 16-19, 2006 held at Cukurova University, Adana, Turkey pp.

Kulkarni, R. M., Naik, D. M., Kachave, D.B., Shinde, S.J. and Jature, S.D., 2009, Effect of Different Chemical Sprays on the Fruit Yield and Quality of Pomegranate (*Punica granatum L*). Var. Ganesh, 2nd International Symposium on "Pomegranate and Minor including Mediterranean Fruits" from June 23-27, 2009 held at UAS, Dharwad, Karnataka, India, pp.111.

Kulkarni, S. R. and Dethe, M. D., 2009, Seasonal Incidence of pests of Pomegranate, 2nd International Symposium on "Pomegranate and Minor including Mediterranean Fruits" from June 23-27, 2009 held at UAS, Dharwad, Karnataka, India, pp.148.

Kulkarni, S. R., Dethe, M. D., and Kale, V.D., 2009, Efficacy of Different insecticides for control of Thrips *Scirtopthrips dorsalis* Hood and their Residues in Pomegranate, 2nd International Symposium on "Pomegranate and Minor including Mediterranean Fruits" from June 23-27, 2009 held at UAS, Dharwad, Karnataka, India, pp.150.

Kulkarni, S. R., Masalkar, S.D., Kadam, J.R. and Chandele, A.G., 2009, Bio-control of Mealy Bugs by *Verticillium lecanii* (Phule Bugicide) on Pomegranate, 2nd International Symposium on "Pomegranate and Minor including Mediterranean Fruits" from June 23-27, 2009 held at UAS, Dharwad, Karnataka, India, pp.152.

Kulkarni, S.R., Dethe, M.D. and Kale, V.D., 2009, Efficacy of Different Insecticides for the control of Fruit Borer *Deudorix isocrates* Fabricius and their residues in Pomegranate, 2nd International Symposium on "Pomegranate and Minor including Mediterranean Fruits" from June 23-27, 2009 held at UAS, Dharwad, Karnataka, India, pp.149.

Kumar, J. and Khosla, K., 2006, Evaluation of pomegranate cultivars in Mid-Hill zone of Himachal Pradesh, India, I[st] International Symposium on "Pomegranate and Minor Mediterranean Fruits" from October 16-19, 2006 held at Cukurova University, Adana, Turkey pp.5.

Kumar, K. and S.N. Lal. 1983. Studies on the Biology, Seasonal Abundance and Host-Parasite Relationship of Fruit Sucking Moth *Othreis fullonia* (Clerk) in Fiji. Fiji Agric. J. 45(2): 71-77.

Kumar, R., Jahagirdar, S., Yenjerappa, S. T. and Patil, H. B., 2006, Epidemiology and management of bacterial blight of pomegranate caused by *Xanthomonas axonopodis* pv. *Punicae*, I[st] International Symposium on "Pomegranate and Minor Mediterranean Fruits" from October 16-19, 2006 held at Cukurova University, Adana, Turkey pp.92.

Labuda, R., 2006, A newly recognized penicillium pathogen discovered on imported pomegranate fruits, I[st] International Symposium on "Pomegranate and Minor Mediterranean Fruits" from October 16-19, 2006 held at Cukurova University, Adana, Turkey pp.94.

Lamalfa, S., Gentile, A., Domina, F. and Tribulato, E., 2006, Primosole: A new selection of Sicilian pomegranate germplasm, I[st] International Symposium on "Pomegranate and Minor Mediterranean Fruits" October 16-19, 2006 held at Cukurova University, Adana, Turkey pp.28.

Levin, G.M., 2006, Pomegranate, published by Third Millennium Published 1931 East Libra Drive Tempe, AZ 85283

Mahadeviah, M., 2009, Technology of Value Added Products of Pomegranate, 2nd International Symposium on "Pomegranate and Minor including Mediterranean Fruits" from June 23-27, 2009 held at UAS, Dharwad, Karnataka, India, pp.123.

Mahmoodi Tabar, S., Davarynejad, G.H., Tehranifar, A. and Nemati, S.H., 2009, Pre-harvest Application of Salicylic Acid and Calcium to improve Qualitative characteristics of Pomegranate Fruit Cv. Shishe Kabe Ferdows, 2nd International Symposium on "Pomegranate and Minor including Mediterranean Fruits" from June 23-27, 2009 held at UAS, Dharwad, Karnataka, India, pp.112.

Mahmoodi Tabar, S., Nemati, S.H., Davarynejad, G.H., Tehranifar, A., 2009, Aril Paleness : New Physiological Disorder That Affect Physical, Chemical, Anatomical and Mineral Nutrient properties of Pomegranate Fruit, 2nd International Symposium on "Pomegranate and Minor including Mediterranean Fruits" from June 23-27, 2009 held at UAS, Dharwad, Karnataka, India, pp.102.

Majid Safa and Javad, K., 2006, Determining the effect of some parameters on extracting the pomegranate arils using a pneumatic method, I[st] International Symposium on "Pomegranate and Minor Mediterranean Fruits" from October 16-19, 2006 held at Cukurova University, Adana, Turkey pp.139.

Mallesh, S.B., Lingaraju, S., Benagi, V.I., Basamma, K., Kumari and Vinaya, H., 2009, Native Plant Growth promoting Rhizobacteria for the suppression of pathogens of Pomegranate (*Punicae granatum L*), 2nd International Symposium on "Pomegranate and Minor including Mediterranean Fruits" from June 23-27, 2009 held at UAS, Dharwad, Karnataka, India, pp.142.

Manjula, N. and Sheikh, M.K., 2008, Knowledge level of Farmers in Management of Bacterial Blight in Pomegranate in Northern Karnataka, Dalimb Virut Samarnika, pp 123-126.

Manjula, N., Sheikh, M.K. and Kunnal, L.B., 2009, Constraints Faced and Suggestions Given by Pomegranate Growers of Bijapur District, 2nd International Symposium on "Pomegranate and Minor including Mediterranean Fruits" from June 23-27, 2009 held at UAS, Dharwad, Karnataka, India, pp.156.

Manjula, N., Sheikh, M.K. and Teggi, M.Y., 200o, Management practices of Bacterial Blight Adopted by Farmers, 2nd International Symposium on "Pomegranate and Minor including Mediterranean Fruits" from June 23-27, 2009 held at UAS, Dharwad, Karnataka, India, pp.155.

Martinez, J.J., Melgarejo, P., Hernandez., F., Legua, P. and Martinez – Font, R., 2006, Polen – pistil compatability of four new pomegranate clones (*Punica granatum L*), Ist International Symposium on "Pomegranate and Minor Mediterranean Fruits" from October 16-19, 2006 held at Cukurova University, Adana, Turkey pp.46.

Masalkar, S.D., Joshi, V.R., Masalkar, S.D., Kulkarni, S.R. and Chavan, S.D., 2009, Effect of different pruning level on yield and quality of Pomegranate, 2nd International Symposium on "Pomegranate and Minor including Mediterranean Fruits" from June 23-27, 2009 held at UAS, Dharwad, Karnataka, India, pp.101.

Masalkar, S.D., Kad, V.P., Kshirsagar, D.B., and Ranpise, S.A., 2009, Effect of various Post Harvest Treatments on Extending Shelf Life of Pomegranate Fruits (Cv. Ganesh), 2nd International Symposium on "Pomegranate and Minor including Mediterranean Fruits" from June 23-27, 2009 held at UAS, Dharwad, Karnataka, India, pp.126.

Matapathi, S. S., Patil, A.B., Jones Nirmalnath, Sheikh, M. K. and Patil, S.R., 2006, Screening of pomegranate (*Punica granatum L.*) cultivars for wine production, Ist International Symposium on "Pomegranate and Minor Mediterranean Fruits" from October 16-19, 2006 held at Cukurova University, Adana, Turkey pp.140.

Matuskovic, J. and Micudova, O., 2006, Practices with chemical mutagen (*Natrium azide*) on growth habit of *Punica granatum L.*, Ist International Symposium on "Pomegranate and Minor Mediterranean Fruits" from October 16-19, 2006 held at Cukurova University, Adana, Turkey pp.41.

Meena, C.L., Dashora, L.K., Prerak Bhatnagar and Virendra Singh, 2009, Integrated Nutrient Management in Pomegranate Cv. Ganesh, 2nd International Symposium on "Pomegranate and Minor including Mediterranean Fruits" from June 23-27, 2009 held at UAS, Dharwad, Karnataka, India, pp.107.

Mehetre, S.S., Masalkar, S.D. and Satbhai, R.D., 2008, Post Harvest Handling of Pomegranate Fruits for Export, Dalimb Virut Samarnika, pp 150-154.

Melgarejo, P., Legua, P., Martinez – Font, R., Martinez, J.J. and Hernandez, F., 2006, Anthocyanin content and colour development of pomegranate jam, Ist International Symposium on "Pomegranate and Minor Mediterranean Fruits" from October 16-19, 2006 held at Cukurova University, Adana, Turkey pp.138.

Melgarejo, P., Legua, P., Martinez – Font, R., Martinez, J.J. and Hernandez, F., 2006, Pomegranate jam preparation, Ist International Symposium on "Pomegranate and Minor Mediterranean Fruits" from October 16-19, 2006 held at Cukurova University, Adana, Turkey pp.141.

Meshram, D.T., Mittal, M.K., Gorantiwar, S.D. and Purohit, R.C., 2009, Water Requirement of Pomegranate (*Punica granatum L*) for Major Growing District (Solapur) for Maharashtra State, 2nd International Symposium on "Pomegranate and Minor including Mediterranean Fruits" from June 23-27, 2009 held at UAS, Dharwad, Karnataka, India, pp.105.

Miguel, M. G., Dandlen, S. and Neves, M.A., 2006, Antioxidant activities of flower extract and its effect on health, I[st] International Symposium on "Pomegranate and Minor Mediterranean Fruits" from October 16-19, 2006 held at Cukurova University, Adana, Turkey pp.86.

Mirdehgham, S. H. and Rahemi, M., 2006, Growth and development of pomegranate fruit. Physical and chemical changes, Ist International Symposium on "Pomegranate and Minor Mediterranean Fruits" from October 16-19, 2006 held at Cukurova University, Adana, Turkey pp.40.

Moghadam, E.G. and Nikkhah, S.H, 2006, Effects of heat treatments of packaging on reduction decay and maintaining quality of pomegranate in storage, I[st] International Symposium on "Pomegranate and Minor Mediterranean Fruits" from October 16-19, 2006 held at Cukurova University, Adana, Turkey pp.105.

Mogle, T.R., Magar, S.J., Suryawanshi, A.P. and Somwanshi, S.D., 2009, Pathogeneicity, Morphology and Cultural characteristics of *Xanthomonas axonopodis Pv.Punicae*, causing only spot of Pomegranate, 2nd International Symposium on "Pomegranate and Minor including Mediterranean Fruits" from June 23-27, 2009 held at UAS, Dharwad, Karnataka, India, pp.137.

Mohseni, A., 2006, The situation of pomegranate orchards in Iran, I[st] International Symposium on "Pomegranate and Minor Mediterranean Fruits" October 16-19, 2006 held at Cukurova University, Adana, Turkey pp.3.

More, H.G., Kad, V.P., Masalkar, S.D., and Kshirasagar, D.B., 2009, Preparation of Carbonated Beverage from Pomegranate Fruit (Cv. Mridula), 2nd International Symposium on "Pomegranate and Minor including Mediterranean Fruits" from June 23-27, 2009 held at UAS, Dharwad, Karnataka, India, pp.128.

Mosse-Robinson, I. 1968. Fruit-Sucking Moths (Lepidoptera: Noctuidae). Australian Zoologist. 14(3): 290-293.

Mulla, S.M., Suryawanshi, A.P., Dhutraj, D.N. and Magar, S.J., 2009, Survey and Survelliance of Bacterial Blight (oily spot) disease of Pomegranata in Maharashtra (India), 2nd International Symposium on "Pomegranate and Minor including Mediterranean Fruits" from June 23-27, 2009 held at UAS, Dharwad, Karnataka, India, pp.136.

Muradoglu, F., Ozrenk, K. and Balta, M.F. Ozturkicu.C., 2006, Pomological identification of pomegranate *(Punica granatum L)* selections from cukurca (Hakkeri), I[st] International Symposium on "Pomegranate and Minor Mediterranean Fruits" from October 16-19, 2006 held at Cukurova University, Adana, Turkey pp.25.

Nabigol, A. and Talaei, A., 2006, The maintenance of pomegranate genetic resources and its breeding in Iran, I[st] International Symposium on "Pomegranate and Minor

Mediterranean Fruits" from October 16-19, 2006 held at Cukurova University, Adana, Turkey pp.23.

Nasri, M., 2006, Pomegranate Geographical habitats and varieties and their ecological needs in Iran, I[st] International Symposium on "Pomegranate and Minor Mediterranean Fruits" from October 16-19, 2006 held at Cukurova University, Adana, Turkey pp.9.

Navale, A.M., Raghuwanshi, K.S., Thakare, C.S. and Borker, S.G., 2009, Chemical control of Fungal leaf and Fruit spot Diseases of Pomegranate, 2nd International Symposium on "Pomegranate and Minor including Mediterranean Fruits" from June 23-27, 2009 held at UAS, Dharwad, Karnataka, India, pp.141.

Navprem Singh, Dhillon, W.S. and Gill, P.P.S., 2009, Quality Improvement Studies in Pomegranate under Sub Tropics of India, 2nd International Symposium on "Pomegranate and Minor including Mediterranean Fruits" from June 23-27, 2009 held at UAS, Dharwad, Karnataka, India, pp.112.

Nomura, K. 1965. Studies on Orchard Illumination Against Fruit-Piercing Moths. II. Some Considerations on the Effect of Orchard Illumination Against Fruit-Piercing Moths. Technical Bulletin of the Faculty of Horticulture of Chiba University. 14: 21-28.

Ozezen, M.Y., 2006, Pomegranate in Turkish culture, Its tree, its flower, its fruit, I[st] International Symposium on "Pomegranate and Minor Mediterranean Fruits" from October 16-19, 2006 held at Cukurova University, Adana, Turkey pp.14.

Ozguven, A.I., Roller, D.W., Yelmaz, C. and Yilmaz, M., 2006, The effects of ecology on the phytoestrogenic composition of Pomegranate (*Punica granatum L*), I[st] International Symposium on "Pomegranate and Minor Mediterranean Fruits" from October 16-19, 2006 held at Cukurova University, Adana, Turkey pp.87.

Ozguven, A.I., Yilmaz, C. and Yilmaz, M., 2006, The situation of pomegranate and minor Mediterranean fruits in Turkiye, I[st] International Symposium on "Pomegranate and Minor Mediterranean Fruits" from, October 16-19, 2006 held at Cukurova University, Adana, Turkey pp.4.

Ozguven, A.I., Yilmaz, C., Rehber, Y. and Yilmaz, M., 2006, The adaptation of different pomegranate varieties in the north Cyprus ecological conditions, I[st] International Symposium on "Pomegranate and Minor Mediterranean Fruits" from October 16-19, 2006 held at Cukurova University, Adana, Turkey pp.37.

Ozturk, N., and Ulusoy, M.R., 2006, Pests and natural enemies determined in Pomegranate orchards in Turkiye, I[st] International Symposium on "Pomegranate and Minor Mediterranean Fruits" from October 16-19, 2006 held at Cukurova University, Adana, Turkey pp.98.

Pala, H., Yilmza, C., Ozguven, A.I. and Tatli, A., 2006, Important disease of pomegranate fruit and control possibilities in Turkiye, I[st] International Symposium on "Pomegranate and Minor Mediterranean Fruits" from October 16-19, 2006 held at Cukurova University, Adana, Turkey pp.101.

Palma, A., Schirra, M., D'Aquino, S., Lamalfa, S., and Continella, G., 2006, Chemical properties changes in pomegranate seeds packaged in polypropylene trays, I[st] International Symposium on "Pomegranate and Minor Mediterranean Fruits" from

from October 16-19, 2006 held at Cukurova University, Adana, Turkey pp.117.

Palou, L. and Del Rio, M.A., 2006, Assessment of fungal pathogens causing post harvest decay of pomegranate in Southeast Spain, I[st] International Symposium on "Pomegranate and Minor Mediterranean Fruits" from October 16-19, 2006 held at Cukurova University, Adana, Turkey pp.106.

Panah, M.H., 2006, The study of pomegranate inspissated juice production with traditional and industrial method and its quality and productivity management, I[st] International Symposium on "Pomegranate and Minor Mediterranean Fruits" from October 16-19, 2006 held at Cukurova University, Adana, Turkey pp.134.

Panahi, M. and Amiri, M. E., 2006, Influence of irrigation and mineral nutrients on growth, yield quantity and quality of pomegranate fruit, I[st] International Symposium on "Pomegranate and Minor Mediterranean Fruits" from October 16-19, 2006 held at Cukurova University, Adana, Turkey pp.55.

Patel, N.M., Patil, D.K., Verma, L.R. and Patel, M.M., 2009, Irrigation Management Through Different Method of Irrigation on Pomegranate Cv. Ganesh, 2nd International Symposium on "Pomegranate and Minor including Mediterranean Fruits" from June 23-27, 2009 held at UAS, Dharwad, Karnataka, India, pp.105.

Patil, A. B., Geeta Goudar and Prema Patil.,2008, Value added products of Pomegranate, Dalimb Virut Samarnika, pp 146-148.

Patil, A. B., Geeta Goudar and Prema, Patil., 2008, Pomegranate : The Nutritional Power House, Dalimb Virut Samarnika, pp 143-145.

Prakash Kulkarni, 2008, Integrated Bacterial Nodal Blight Management of Pomegranate, Dalimb Virut Samarnika, pp 127-128.

Prasad, R.N., 2009, Effect of planting Density on Growth, Yield and Quality of Pomegranate under Arid conditions of Western Rajasthan, 2nd International Symposium on "Pomegranate and Minor including Mediterranean Fruits" from June 23-27, 2009 held at UAS, Dharwad, Karnataka, India, pp.104.

Prashanth, A. and Sataraddi, A.R., 2009, Variability in Different Isolates of Anthracnose of Pomegranate caused by *Colletotrichum gloeosporioides*, 2nd International Symposium on "Pomegranate and Minor including Mediterranean Fruits" from June 23-27, 2009 held at UAS, Dharwad, Karnataka, India, pp.142.

Proboglav, E.S., Aruthaynan, L.G., Mukhamedkanov, T. and Perova, L.P., 1983, Manufacture of apple-based Acerry Type Wine, International Journal for Food Science and Technology 15:1211-1666.

Pulla Reddy, V., Reddy, Y.N., Narayanareddy, P and Chandrasekhar, R., 2009, Effect of Precursor Microbial Interactions on Induction of Flowering and Fruits in Pomegranate, 2nd International Symposium on "Pomegranate and Minor including Mediterranean Fruits" from June 23-27, 2009 held at UAS, Dharwad, Karnataka, India, pp.99.

Pulla reddy., Ch., Reddy, Y.N., Narayanareddy, P. and Chandrasekhar, R., 2009, Effect of Ethylene inhibiting chemicals on yield and quality of Pomegranate, 2nd International Symposium on "Pomegranate and Minor including Mediterranean Fruits" from June 23-27, 2009 held at UAS, Dharwad, Karnataka, India, pp.109.

Raghuvanshi, K.S. and Masalkar, S.D., 2008, Management of Pomegranate Wilt, Dalimb Virut Samarnika, pp 141-142.

Raghuwanshi, K.S., Borkar, S.G. and Sonawane, R.B., 2009, In vitro Evaluation of pesticides against *Xanthomonas axonopodis Pv.Punicae*, 2nd International Symposium on "Pomegranate and Minor including Mediterranean Fruits" from June 23-27, 2009 held at UAS, Dharwad, Karnataka, India, pp.138.

Rahim, M.A. and Huq, S.M., 2006, Pomegranate – a promising crop of Bangladesh, I[st] International Symposium on "Pomegranate and Minor Mediterranean Fruits" from October 16-19, 2006 held at Cukurova University, Adana, Turkey pp.1.

Rahim, M.A., and Huq, S.M., 2006, Integrated crop management for controlling pomegranate fruit borer and powdery mildew, I[st] International Symposium on "Pomegranate and Minor Mediterranean Fruits" October 16-19, 2006 held at Cukurova University, Adana, Turkey pp.96.

Raja, M. E., 2006, Important micronutrient disorders in Pomegranate in different soil types of India, I[st] International Symposium on "Pomegranate and Minor Mediterranean Fruits" from October 16-19, 2006 held at Cukurova University, Adana, Turkey pp.95.

Raja, M.E., 2006, Studies on susceptibility of Pomegranate cultivars in India to nutrient disorders, I[st] International Symposium on "Pomegranate and Minor Mediterranean Fruits" from October 16-19, 2006 held at Cukurova University, Adana, Turkey pp.89.

Ram Asrey., 2006, Pomegranate Post Harvest Management in India, I[st] International Symposium on "Pomegranate and Minor Mediterranean Fruits" from October 16-19, 2006 held at Cukurova University, Adana, Turkey pp.103.

Ramachandra, Jadhav, C.T., Jyotsana Sharma and Marathe, R.A., 2009. Effect of grafting methods and time of scion sprouting, graft success and subsequent growth on grafted plants of pomegranate (*Punica granatum L*). Cv. Bhagwa, 2nd International Symposium on "Pomegranate and Minor including Mediterranean Fruits" from June 23-27, 2009 held at UAS, Dharwad, Karnataka, India, pp.69.

Ramanjni Gowda, P.H., 2009, Biotechnological Intervention for the improvement of pomegranate and Minor fruits, 2nd International Symposium on "Pomegranate and Minor including Mediterranean Fruits" from June 23-27, 2009 held at UAS, Dharwad, Karnataka, India, pp.69.

Ramchandra, Jadhav, V.T., Dhinesh Babu, K. and Maity, A., 2009, Influence of Chemical Defoliants on Defoliation and Twig Sprouting in Pomegranate (*Punica granatum L*) Cv.Bhagwa, 2nd International Symposium on "Pomegranate and Minor including Mediterranean Fruits" from June 23-27, 2009 held at UAS, Dharwad, Karnataka, India, pp.110.

Ranjbar, H., Sarkhosh, A. and Ghasemmegad, M., 2006, The Effects of Calcium chloride Hot Water Treatment and Polyethylene Bag Packaging on the Storage Life and Quality of Pomegranate (cv: Malas – Saveh), I[st] International Symposium on "Pomegranate and Minor Mediterranean Fruits" from October 16-19, 2006 held at Cukurova University, Adana, Turkey pp.113.

Ranpise, S.A., Masalkar, S.D., Kad, V.P., Kshirsagar, D.B., 2009, Effect of pre-harvest sprays of chemicals for improving the quality of pomegranate fruits (*Punica granatum L*), 2nd International Symposium on "Pomegranate and Minor including Mediterranean Fruits" from June 23-27, 2009 held at UAS, Dharwad, Karnataka, India, pp.111.

Ravikumar, K.Y. and Hosamani, S.B., 2009, Export performance of pomegranate in India, 2nd International Symposium on "Pomegranate and Minor including Mediterranean Fruits" from June 23-27, 2009 held at UAS, Dharwad, Karnataka, India, pp.157.

Ravikumar, M.R., Wali, S.Y., Benagi, V.I., Patil, H.B. and Suresh, S. Patil, 2009, Management of Bacterial Blight of Pomegranate through Chemical/Antibiotics, 2nd International Symposium on "Pomegranate and Minor including Mediterranean Fruits" from June 23-27, 2009 held at UAS, Dharwad, Karnataka, India, pp.132.

Ravikumar, M.R., Yenjerappa, S.T., Benagi, V.I., Suresh.S.Patil., Patil, H.B. and Wali, S.Y., Integrated Management of Bacterial Blight of Pomegranate caused by *Xanthomonas axonopodis Pv. Punicae*, 2nd International Symposium on "Pomegranate and Minor including Mediterranean Fruits" from June 23-27, 2009 held at UAS, Dharwad, Karnataka, India, pp.139.

Ravindran, C., Guptal, S.P., Ravikher, and Sharma, M., 2006, Minor fruits for food – Nutrition Security and Enhanced Rural livelihood for Tribal Communities of Intermediate Zone of Jammu and Kashmir, Ist International Symposium on "Pomegranate and Minor Mediterranean Fruits" from October 16-19, 2006 held at Cukurova University, Adana, Turkey pp.7.

Ravindran, C., Jaganath, S. and Krishna Manohar, R., 2009, Economics of Propagation of Pomegranate (*Punica granatum L*) under Protected Environmental Conditions, 2nd International Symposium on "Pomegranate and Minor including Mediterranean Fruits" from June 23-27, 2009 held at UAS, Dharwad, Karnataka, India, pp.154.

Ravindra, R., 2009., APEDAs Role in Enhancing the Exports of Pomegranate, 2nd International Symposium on "Pomegranate and Minor including Mediterranean Fruits" from June 23-27, 2009 held at UAS, Dharwad, Karnataka, India, pp.153.

Rezaian, S., Zabihi, H.R. and Atta Hosseini, S.A., 2006, Recognition of nutritional reasons for pomegranate colorless seed formation in the Ferdows region of Khorasan province in Iran, Ist International Symposium on "Pomegranate and Minor Mediterranean Fruits" from October 16-19, 2006 held at Cukurova University, Adana, Turkey pp.42.

Richard, A., Barbara, B. and David, S., 2006, The Incredible Pomegranate Plant and Fruit, Published 1931 East Libra Drive Tempe, AZ 85283,.

Rokhade, A.K., Sowjanya, G., Patil, C.P. and Swamy, G.S.K., 2009, Preparation of Anardana from Pomegranate, 2nd International Symposium on "Pomegranate and Minor including Mediterranean Fruits" from June 23-27, 2009 held at UAS, Dharwad, Karnataka, India, pp.125.

Rokhade, A.K., Sowjanya, G., Patil, C.P., Madalageri, M.B., 2009, Preparation and Storage of Carbonated Ready to Serve (RTS) Pomegranate Beverage, 2nd International Symposium on "Pomegranate and Minor including Mediterranean Fruits" from June 23-27, 2009 held at UAS, Dharwad, Karnataka, India, pp.127.

Rouhani, S. and Tahere, H., 2006, Chemo preventive Activity of Different Colorant Fractions Separated from Two kind of Iranian Pomegranate rind, I[st] International Symposium on "Pomegranate and Minor Mediterranean Fruits" from October 16-19, 2006 held at Cukurova University, Adana, Turkey pp.115.

Rouhani, S., Shafaie, S.Z. and Alami, R., 2006, Influence of Different Ingredients on the Extraction and Colour Stability of Iranian Pomegranate Rind Colorants, I[st] International Symposium on "Pomegranate and Minor Mediterranean Fruits" from October 16-19, 2006 held at Cukurova University, Adana, Turkey pp.116.

Rymon, D., 2006, Intercountry compositive economic analysis of pomegranate production cost, I[st] International Symposium on "Pomegranate and Minor Mediterranean Fruits" from October 16-19, 2006 held at Cukurova University, Adana, Turkey pp.122.

Sachs, Y., Ward, G., Agar, T. and R.Porat, 2006, Preserving pomegranate quality in modified atmosphere packaging, I[st] International Symposium on "Pomegranate and Minor Mediterranean Fruits" October 16-19, 2006 held at Cukurova University, Adana, Turkey.pp.122.

Sadiye GozLekki, Sahriya SONMEZ and Filiz Okturen, 2009, Determination of Macro and Micro Nutrient contents of Pomegranate (*Punica granatum L*) Cv. Hicaznar) Fruits Grown in the west Mediterranean Region of Turkey, 2nd International Symposium on "Pomegranate and Minor including Mediterranean Fruits" from June 23-27, 2009 held at UAS, Dharwad, Karnataka, India, pp.123.

Safa, M., Khazaei, J. and, Kianmehr,M., 2006, Extracting the pomegranate seeds (arils) using a pneumatic method, I[st] International Symposium on "Pomegranate and Minor Mediterranean Fruits" from October 16-19, 2006 held at Cukurova University, Adana, Turkey.pp.126.

Sagar, V. R., 2006, Preparation and storage study of Pomegranate powder, I[st] International Symposium on "Pomegranate and Minor Mediterranean Fruits" from October 16-19, 2006 held at Cukurova University, Adana, Turkey. pp.133.

Salgado, L., Melgarejo, P. and Sanchez., M., 2006, Antimicrobial activity of crude extracts from *Punica granatum L* Fruit, I[st] International Symposium on "Pomegranate and Minor Mediterranean Fruits" from October 16-19, 2006 held at Cukurova University, Adana, Turkey.pp.84.

Sarig,Y., Juran and Galili, A., 2009, Pomegranate post-harvest Technologies Current state of the art, and Beyond, 2nd International Symposium on "Pomegranate and Minor including Mediterranean Fruits" from June 23-27, 2009 held at UAS, Dharwad, Karnataka, India, pp.121.

Santosh, R.Kulkarni, 2008, Pest Management of Pomegranate, Dalimb Virut Samarnika, pp 129-137.

Santosh,R.Kulkarni, 2008, *Verticillium lecanii*: The effective entomopathogen against mealy bugs on pomegranate, Dalimb Virut Samarnika, pp 138-139.

Sarkhosh, A., Zamani, Z. and Fatahi, M.R., 2006, A review on medicinal characteristics of Pomegranate (*Punica granatum L.*), I[st] International Symposium on "Pomegranate and Minor Mediterranean Fruits" October 16-19, 2006 held at Cukurova University, Adana, Turkey.pp.77.

Sarkhosh, A., Zamani, Z., Fatahi, M.R., Ebadi, A., Akrami, M.R. and Tabatabaei, S.Z., 2006, Study on relationships among quantitative and qualitative characteristics of fruit components of pomegranate genotypes, I[st] International Symposium on "Pomegranate and Minor Mediterranean Fruits" from October 16-19, 2006 held at Cukurova University, Adana, Turkey.pp.68.

Sataraddi, A.R., Prashant, A., Virupaksh Prabhu, H., Jamadar, M.M. and Subhash Aski, 2009, Role of Bioagents and Botanicals in the Management of Anthracnose of Pomegranate, 2nd International Symposium on "Pomegranate and Minor including Mediterranean Fruits" from June 23-27, 2009 held at UAS, Dharwad, Karnataka, India, pp.143.

Sepahvand, Kh., Asgari, M. and Saei, A., 2006, Evaluation of the Morphological and fruit characters and cold Hardiness of ten pomegranate (*Punica granatum L*) Accessions cultivated in Karaj, I[st] International Symposium on "Pomegranate and Minor Mediterranean Fruits" from October 16-19, 2006 held at Cukurova University, Adana, Turkey pp.16.

Sharifani, M., Adeh, A. and Moghadan, V.E., 2006, Botanical diversity in wild genetic post and *Punica granatum* in Golestan province, I[st] International Symposium on "Pomegranate and Minor Mediterranean Fruits" from October 16-19, 2006 held at Cukurova University, Adana, Turkey pp.24.

Sharma, B.D., More, T.A., Bhargava, R and Singh, R.S., 2009, Effects of Organic Manures and Inorganica Fertilizers on Growth, Physiological Attributes, Fruit Yield and Quality Parameters of Pomegranate (*Punica granatum L*), 2nd International Symposium on "Pomegranate and Minor including Mediterranean Fruits" from June 23-27, 2009 held at UAS, Dharwad, Karnataka, India, pp.100.

Sharma, K.K., Jyotsana Sharma, Jadhav, V.T., and Ramachandra, 2009, Present status of Pomegrnaate Bacterial Blight caused by *Xanthomonas axonopodis Pv. Punicae* and its Management, 2nd International Symposium on "Pomegranate and Minor including Mediterranean Fruits" from June 23-27, 2009 held at UAS, Dharwad, Karnataka, India, pp.135.

Sharma, M., Ravindran, C., Gupta, S.P. and Ravikher., 2006, Genetic variability studies in wild pomegranate (*Punica protopunica* Bele.F.) in Rajouri and Poonch District of Jammu and Kashmir, I[st] International Symposium on "Pomegranate and Minor Mediterranean Fruits" from October 16-19, 2006 held at Cukurova University, Adana, Turkey pp.17.

Sharma, N. and Chaitanya Belsore; 2009, Effect of Plant Bio-Regulators and Nutrients on Fruit Cracking and Quality of Pomegranate (*Punica granatum L*) Cv.G-137, 2nd International Symposium on "Pomegranate and Minor including Mediterranean Fruits" from June 23-27, 2009 held at UAS, Dharwad, Karnataka, India, pp.109.

Sharma, R.L. and Tegta, R.K., 2009, Incidence of Dry Rot of Pomegranate in Himachal Pradesh and its Management, 2nd International Symposium on "Pomegranate and Minor including Mediterranean Fruits" from June 23-27, 2009 held at UAS, Dharwad, Karnataka, India, pp.133.

Sheikh, M. K. and Rao, M. M., 2006, Effect of chemical on control of fruit cracking in pomegranate (*Punica granatum L*), I[st] International Symposium on "Pomegranate and Minor Mediterranean Fruits" from October 16-19, 2006 held at Cukurova University, Adana, Turkey pp.62.

Sheikh, M. K. and Rao, M. M., 2006, Effect of chemicals on fruit, rind and aril colour in pomegranate (*Punica granatum L*) var. Ganesh, I[st] International Symposium on "Pomegranate and Minor Mediterranean Fruits" from October 16-19, 2006 held at Cukurova University, Adana, Turkey pp.60.

Sheikh, M. K. and Rao, M. M., 2006, Effect of split application of N and K on growth and fruiting in ganesh pomegranate (*Punica granatum L*), I[st] International Symposium on "Pomegranate and Minor Mediterranean Fruits" from October 16-19, 2006 held at Cukurova University, Adana, Turkey pp.61.

Sheikh, M.K. and Manjula, N., 2001, Medicinal Importance of Pomegranate Fruit, Dalimb Virut Samarnika, pp 165-166.

Sheikh, M.K. and Manjula, N., 2009, Control of Fruit Cracking in Pomegranate (*Punica granatum L*), 2nd International Symposium on "Pomegranate and Minor including Mediterranean Fruits" from June 23-27, 2009 held at UAS, Dharwad, Karnataka, India, pp.108.

Sheikh, M.K. and Rao, M.M., 2006, Effect of pruning and fruit load on yield and quality in pomegranate, I[st] International Symposium on "Pomegranate and Minor Mediterranean Fruits" from October 16-19, 2006 held at Cukurova University, Adana, Turkey pp.63.

Shivlal, N., Ahmed, Verma, M.K., Dinesh Kumar and Reshi, T.A., 2009, Effect of coating Materials on Shelflife of Pomegranate (*Punica granatum L*) fruit under Ambient storage condition, 2nd International Symposium on "Pomegranate and Minor including Mediterranean Fruits" from June 23-27, 2009 held at UAS, Dharwad, Karnataka, India, pp.124.

Shlomo Marco., 2009, Plant Protection in High Yielding Pomegranate Plantations – present situation and prospects for the future, 2nd International Symposium on "Pomegranate and Minor including Mediterranean Fruits" from June 23-27, 2009 held at UAS, Dharwad, Karnataka, India, pp.146.

Shmilovich, Z., Sarig, Y., Ronen, B., Offman, A., Egozi, H., Beres, H., Bar-lev, E. and Groz, F., 2006, Development of a method and a system for extracting the seeds (arils) from Pomegranate fruits from concept to commercial utilization, I[st] International Symposium on "Pomegranate and Minor Mediterranean Fruits" from October 16-19, 2006 held at Cukurova University, Adana, Turkey pp.127.

Shukla, M. Gupta, K., Khan, K.A. and Maqqi, T. M., 2006, Prevention of human cartilage degradation by Pomegranate, I[st] International Symposium on "Pomegranate and Minor Mediterranean Fruits" from October 16-19, 2006 held at Cukurova University, Adana, Turkey pp.79.

Singh, A.B., 2009, Effect of Organic Nutrient Sources on Fruit Yield on Quality of Pomegranate, 2nd International Symposium on "Pomegranate and Minor including

Mediterranean Fruits" from June 23-27, 2009 held at UAS, Dharwad, Karnataka, India, pp.100.

Singh, D.B., Gupta, R.K. and Patil, R.T., 2009, Effect of Minimal Processing of Pomegranate Arils on Quality Maintenance and Shelf Life, 2nd International Symposium on "Pomegranate and Minor including Mediterranean Fruits" from June 23-27, 2009 held at UAS, Dharwad, Karnataka, India, pp.121.

Singh, D.B., Gupta, R.K. and Patil, R.T., 2009, Quality of Shelf life of Pomegranate Fruits as Affected by Post Harvest Treatments, packaging materials and storage conditions, 2nd International Symposium on "Pomegranate and Minor including Mediterranean Fruits" from June 23-27, 2009 held at UAS, Dharwad, Karnataka, India, pp.126.

Singh, D.B., Kingsly, A.R.P and Wanjari, O.D., 2006, Effect of Drying Temperature on Quality Retention of Dehydrated Pomegranate Arils (Anardana), Ist International Symposium on "Pomegranate and Minor Mediterranean Fruits" from October 16-19, 2006 held at Cukurova University, Adana, Turkey pp.111.

Singh, D.B., Samadia, D.K. and Kingsly, A.R.P., 2006, Conservation, characterization and evaluation of pomegranate germplasm under Arid ecosystem of India, Ist International Symposium on "Pomegranate and Minor Mediterranean Fruits" from October 16-19, 2006 held at Cukurova University, Adana, Turkey pp.15.

Singhnagi, H .P. and Manjrekar S.P., 1975 Studies on the Preparation of Cider from North Indian Apples: II Storage Studies Indian Food Packer, 29:12-15

Somasekhara, Y.M. and Gaddanakeri, M.A., 2009, Host specificity of Pomegranate (*Punica granatum L*) wilt pathogen, *Ceratocystis fimbriate*, 2nd International Symposium on "Pomegranate and Minor including Mediterranean Fruits" from June 23-27, 2009 held at UAS, Dharwad, Karnataka, India, pp.139.

Somasekhara, Y.M., 2009, In vitro and In vivo Evaluation of Fungicides for the Management of Pomegranate wilt pathogen (*Ceratocystis fimbriate L*), 2nd International Symposium on "Pomegranate and Minor including Mediterranean Fruits" from June 23-27, 2009 held at UAS, Dharwad, Karnataka, India, pp.140.

Somasekhara, Y.M., Ravichandra, N.G. and Jain, A.K., 2009, Biological Management of Root Knot Nematode (*Meloidogyne Incognita*) infesting Pomegranate (*Punica granatum L*) in Karanataka, 2nd International Symposium on "Pomegranate and Minor including Mediterranean Fruits" from June 23-27, 2009 held at UAS, Dharwad, Karnataka, India, pp.143.

Somasekhara, Y.M., Wali, S.Y. and Sheikh, M.K., 2009, Studies and the management of Pomegranate (*Punica granatum L*) wilt (*Ceratocystis fimbriata*, Latin Americal Group), 2nd International Symposium on "Pomegranate and Minor including Mediterranean Fruits" from June 23-27, 2009 held at UAS, Dharwad, Karnataka, India, pp.132.

Sreedevi, K. and Abraham Verghese., 2009, Effect of Aphid, *Aphis punicae passerine* (Hemiptera : Aphididae) infectation on Pomegranate Fruits and Flowers, 2nd International Symposium on "Pomegranate and Minor including Mediterranean Fruits" from June 23-27, 2009 held at UAS, Dharwad, Karnataka India, pp.150.

Subramanyam, K., 2009, Pomegranate Bacterial Blight Disease Situation in Andhra Pradesh,

2nd International Symposium on "Pomegranate and Minor including Mediterranean Fruits" from June 23-27, 2009 held at UAS, Dharwad, Karnataka, India, pp.136.

Suresh, E. R. and Ethir. Aj. S. 1988, Studies on the Preparation and Quality Improvement of Wine. Annual Report, Indian Insititute of Horticultural Research Bangalore, P.53.

Suresh, E.R., Ethiraj. S. and Negi, S.S., 1985 Evaluation of New Grape Cultivars for Prepration of Wine. Journal of Food Science and Technology, 22:211-215

Suryawanshi, K.S., More,T.R.,Pilare,V.D., and Somwanshi, S.D., 2009, Efficacy of Antibiotics and Fungicides against oily spot disease of Pomegranate, 2nd International Symposium on "Pomegranate and Minor including Mediterranean Fruits" from June 23-27, 2009 held at UAS, Dharwad, Karnataka, India, pp.137.

Swati Barche, Kamal Singh Kirad and Singh, D.B., 2009, Plant Growth Regulators and Fungicide on Rooting of Harwood cuttings in Pomegranate, 2nd International Symposium on "Pomegranate and Minor including Mediterranean Fruits" from June 23-27, 2009 held at UAS, Dharwad, Karnataka, India, pp.71.

Tabatabaei, S.Z. and Sarkhosh, A., 2006, Analysis and comparison of salinity tolerance among 10 Iranian pomegranate commercial cultivars, Ist International Symposium on "Pomegranate and Minor Mediterranean Fruits" from October 16-19, 2006 held at Cukurova University, Adana, Turkey pp.54.

Tabatabaei, S.Z. and Sarkhosh, A., 2006, Soil application of paclobutrazol and uniconazol on yield component of pomegranate (*Punica granatum L*) cvs. Shavar and Zagh and morphological studies of flower in Yazd province, Ist International Symposium on "Pomegranate and Minor Mediterranean Fruits" from October 16-19, 2006 held at Cukurova University, Adana, Turkey pp.49.

Tabatebaei, S.Z., and Sarkhosh, A., 2006, Analysis and demonstrated the most resistance and the most sensitive pomegranate genotypes to cracking and carbo moth in Saveh Pomegranate collection, Ist International Symposium on "Pomegranate and Minor Mediterranean Fruits" from October 16-19, 2006 held at Cukurova University, Adana, Turkey pp.90.

Thakur, A.K., Patil, R.T., Singh, D.B. and Gupta, R.K., 2009, Development of Tool and Machine for Safe Separation of Aril from Pomegranate, 2nd International Symposium on "Pomegranate and Minor including Mediterranean Fruits" from June 23-27, 2009 held at UAS, Dharwad, Karnataka, India, pp.124.

Tripler, E., Harari, M. and Borochov – Neori, M., 2006, Effects of heat and radiation accumulation during maturation and ripening on pomegranate fruit quality, Ist International Symposium on "Pomegranate and Minor Mediterranean Fruits" from October 16-19, 2006 held at Cukurova University, Adana, Turkey pp.43.

Tryon, H. 1898. Orange-Piercing Moths - Fam. Ophiderinae. Queensland Agric. J. 2: 308-315.

Tumer, L., Ozguven, A. I. and Tumer, M. A., 2006, Changes on the maturation stages of some pomegranate varieties, Ist International Symposium on "Pomegranate and Minor Mediterranean Fruits" from October 16-19, 2006 held at Cukurova University, Adana, Turkey pp.58.

Varasteh, F., Arzani, K., Zamani, Z. and Mohseni, A., 2006, Evaluation of the most important fruit characteristics of some commercial pomegranate (*Punica granatum L*) cultivars of Iran, I[st] International Symposium on "Pomegranate and Minor Mediterranean Fruits" from October 16-19, 2006 held at Cukurova University, Adana, Turkey pp.20.

Vasanth Kumar, G.K., 2006, Pomegranate cultivation in Karnataka State, India – A profitable venture, I[st] International Symposium on "Pomegranate and Minor Mediterranean Fruits" from October 16-19, 2006 held at Cukurova University, Adana, Turkey pp.6.

Vijay Bahadur, Hota, R.K., Devi Singh and Singh, D.B., 2009, Effect of Rooting Media and Indole Butyric Acid on Rooting and Survival of cuttings in Pomegranate (*Punica granatum L*). 2nd International Symposium on "Pomegranate and Minor including Mediterranean Fruits" from June 23-27, 2009 held at UAS, Dharwad, Karnataka, India, pp.73.

Vijay Kolekar., 2009, Monitoring of the Pesticide Residues in Exportable Pomegranate Fruits, 2nd International Symposium on "Pomegranate and Minor including Mediterranean Fruits" from June 23-27, 2009 held at UAS, Dharwad, Karnataka, India, pp.145.

Vardin, H. and Fenercioglu, H., 2006, A study on the development of pomegranate juice processing technology : Pressing of pomegranate fruits, I[st] International Symposium on "Pomegranate and Minor Mediterranean Fruits" from October 16-19, 2006 held at Cukurova University, Adana, Turkey pp.132.

Wang, R., Xiang, L., Wang, W., Xing, D., and Du, L., 2006, Antioxidant activity of the seeds and flowers of *Punica granatum*, I[st] International Symposium on "Pomegranate and Minor Mediterranean Fruits" from October 16-19, 2006 held at Cukurova University, Adana, Turkey pp.83.

Warade, S.D., Joshi, V.R., Masalkar, S.D. and Kulkarni, S.R., 2009, Effect of Different Mulches on Yield and Quality of Pomegranate, 2nd International Symposium on "Pomegranate and Minor including Mediterranean Fruits" from June 23-27, 2009 held at UAS, Dharwad, Karnataka, India, pp.123.

Waterhouse, D.F. and K.R. Norris. 1987. Chapter 29 *Othreis fullonia* (Clerk). pp. 240-249. In: Biological Control Pacific Prospects. Inkata Press; Melbourne. 454 pages.

Whitehead, V.B. and Rust, D.J. 1967. Orchard Illumination as a Counter to the Fruit Piercing Moths. Decid. Fruit Grow. 17: 357-358.

Whitehead, V.B. and Rust, D.J. 1972. Fruit-Piercing Moths: Ecology and Reduction of Damage. Dried Fruit 4: 4 pp.

Wohlfarter, M., Gilliomee, J.H. and Venter, E., 2009, Nanophyes spp. Recorded from Commercial Pomegranate, *Punica granatum*, in Western Cape Province of South Africa, 2nd International Symposium on "Pomegranate and Minor including Mediterranean Fruits" October 16-19, 2006 held at Cukurova University, Adana, Turkey pp.149.

Xiang, L., Xing, D., Xu.L., Din, Y., Wang, W. and Du, L., 2006, Determination of ellagic acid in pomegranate leaves : Influence of varieties, seasons and processing methods, I[st] International Symposium on "Pomegranate and Minor Mediterranean Fruits" from October 16-19, 2006 held at Cukurova University, Adana, Turkey pp.48.

Yahya Selahvarzi, Ali Tehranifar, Gholamhossein Davarinejed and Hossein Nimate, 2009, Bio-index study of Pomegranate (*Punica granatum L*) chilling injury, 2nd International Symposium on "Pomegranate and Minor including Mediterranean Fruits" from June 23-27, 2009 held at UAS, Dharwad, Karnataka, India, pp.113.

Yahya Selahvarzi, Ali Tehranifar, Vahid Jahanbakhsh, 2009, Antifungal Activitiy of Aqueous and Methanolic Extracts of Different parts of Pomegranate Tree on some Post Harvest Fungi, 2nd International Symposium on "Pomegranate and Minor including Mediterranean Fruits" from June 23-27, 2009 held at UAS, Dharwad, Karnataka, India, pp.134.

Yazici, K. and Kayank, L., 2006, Effect of Kaolin and shading treatments on sunburn in fruit Hicaznar cultivar of Pomegranate (*Punica granatum L.* cv. Hicaznar), I[st] International Symposium on "Pomegranate and Minor Mediterranean Fruits" from October 16-19, 2006 held at Cukurova University, Adana, Turkey pp.39.

Yazici, K. and Kaynak, L., 2006, The effects of air temperature relative humidity and solar radiation on fruit surface temperatures and sunburn damage in pomegranate (*Punica granatum L* cv. Hicaznar), I[st] International Symposium on "Pomegranate and Minor Mediterranean Fruits" from October 16-19, 2006 held at Cukurova University, Adana, Turkey pp.47.

Yenjerappa, S.T., Naragund, V.B. and Jawadagi, R.S., 2009, Management of Bacterial Blight of Pomegranate through cultural methods, 2nd International Symposium on "Pomegranate and Minor including Mediterranean Fruits" from June 23-27, 2009 held at UAS, Dharwad, Karnataka, India, pp.132.

Yenjerappa, S.T., Naragund, V.B. and Rafee, C.M., 2009, Field performance of Bactericides, Bioagents and Botanicals against the Bacterial Blight in Pomegranate, 2nd International Symposium on "Pomegranate and Minor including Mediterranean Fruits" from June 23-27, 2009 held at UAS, Dharwad, Karnataka, India, pp.138.

Yildiz, H., Obuz, E. and Bayraletaroglu, G., 2006, Pomegranate: Its antioxidant activity and its effect on health, I[st] International Symposium on "Pomegranate and Minor Mediterranean Fruits" from October 16-19, 2006 held at Cukurova University, Adana, Turkey pp.85.

Yilmaz, C. and Ozguven, A. I., 2006, The effect of some plant nutrients, gibberlic acid and pinolene treatments on the yield, fruit quality and cracking in pomegranate, I[st] International Symposium on "Pomegranate and Minor fruits Mediterranean Fruits" from June 16-19, 2006 held at Cukurova University, Adana, Turkiye, pp.59.

Yilmaz, M., Ake-Kacar, Y., Yalcin – Mendi, Y., and Isik – Ozguven, A., 2006, Molecular characterization of some pomegranate genotypes by RAPD markers, I[st] International Symposium on "Pomegranate and Minor Mediterranean Fruits" from October 16-19, 2006 held at Cukurova University, Adana, Turkey pp.72.

Zamani, Z., Sarkhosh, A., Fatahi, M.R. Ebadi, A., 2006, Evaluation of RAPD markets for assessment of genetic variation between pomegranate genotypes, I[st] International Symposium on "Pomegranate and Minor Mediterranean Fruits" from October 16-19, 2006 held at Cukurova University, Adana, Turkey pp.65.

Zamani, Z., Zarei, A.K., Fatahi Moghadam, M.R., 2009, A wild accession of pomegranate showing high difference from cultivated ones at DNA level, 2nd International Symposium on "Pomegranate and Minor including Mediterranean Fruits" from June 23-27, 2009 held at UAS, Dharwad, Karnataka, India, pp.69-70.

Zolfagharieh, H., Babali, M., Omidpour, A., Mostafavi, H., and Tabatabail, Z., 2009, Evaluation of Potential of Sterile Insect Technique in controlling *Ectomyelois Ceratoniae Zeller*, 2nd International Symposium on "Pomegranate and Minor including Mediterranean Fruits" from June 23-27, 2009 held at UAS, Dharwad, Karnataka, India, pp.146.

Zolfagherieh, H., Omidpour, A., Azmayesh, Fard, P. and Goldansaz., 2006, The laboratory investigation on inherited sterility of Pomegranate fruit moth (*Ectomyeloise ceratoniae*), Ist International Symposium on "Pomegranate and Minor Mediterranean Fruits" October 16-19, 2006 held at Cukurova University, Adana, Turkey pp.88.

Zolfeghari Nasab.R., Hadian, J., Ghasemnegad, M. and Ranjbar, H., 2006, Reducing chilling injury and decay incidence in pomegranate fruits (Malas Saveh) by Methyl Salicylate, Ist International Symposium on "Pomegranate and Minor Mediterranean Fruits" October 16-19, 2006 held at Cukurova University, Adana, Turkey pp.109.

Annexure - I

AEZ - POMEGRANATE

POMEGRANATE EXPORTS FROM INDIA-DURING 1998-99, 1999-2000 & 2000-02

Qty in Kgs. Value in Rs.

Country	1998-99 Quantity	Value	1999-2000	Value	2000-01 Quantity	Value
1	2	3	4	5	6	7
American Samoa	-	-	36	1500		
Australia	-	-	-	-	2000	23493
Belgium	-	-	20800	753480	31200	1027884
Bangladesh	296167	4440822	297506	2529939	140564	224631
Bahrain	194994	35933372	160436	2728963	193174	6421599
Bahamas	-	-	--	18251	475466	
Brazil	20250	1071429	-	-	-	-
Barbodos	-	-	17750	52442	-	
Canada	36583	10200043	60743	1459593	50040	1284685
Spain	108000	2137085	10950	1408755	-	-
France	35050	783401	400	4351	-	-
Germany	48380	1865189	5698	59823	10000	287100
U.K.	272826	10700621	241554	8829223	161122	6867031
Ghana	600	6530	-	-	-	-
Hong kong	15220	94400	1425	71572	-	
Italy	-	-	-	-	140001	1999200
Japan	5450	196084	10200	448014	11000	377308
Kenya	6600	147833	-	-	-	
Kuwait	125247	2817975	139113	1893082	66625	1162297
LaoPd.Dem. Rep	7740	143871	-	-	-	
Srilanka	237549	4689777	254505	3445519	148362	2464700
Maldives	800	10729	264	8108	-	-
Mexico	20000	1327303	-	-	-	-

Malaysia	60000	152640	500	24500	-	-
Netherlands	12590	768450	-	-	8000	1534590
Norway	100	2150	-	-	-	-
Nepal	-	-	212	21800	13000	74952
Oman	26890	11177453	49585	795249	182372	4801151
Pakistan	32210	369649	15787	204363	32204	481447
Phillippines	325494	2345108	-	-	-	-
Portugal	35494	2345107	-	-	-	-
Paraguay	1200	24850	-	-	-	-
Qatar	42240	581683	16110a	189261	12387	322862
Rwanda	3110	123719	-	-	-	-
South Africa	-	-	550	26187	47620	805784
Soudi Aribia	790234	11045771	1233470	18613220	170644	3266862
Singapore	12630	320593	22992	482245	1938	37155
Sweden	4709	100430	-	-	-	-
Thailand	1800	85236	-	-	-	-
Trinidad & Tabago	20250	106507	-	-	-	-
Tanzania	300	11497	-	-	-	-
U.A.E.	1758470	35113168	3159530	69523495	292227	62570885
U.S.A.	27860	521328	1750	31983	14575	119513
Product Total	4239148	89606634	5726366	115368664	445541	99155636

Source: APEDA Export Statistics for Agro & Food Products-2000-01 APEDA Regional Office, Hyderabad.

ANNEXURE - II

AEZ-POMEGFRANATE

Production, Export of Pomegranate in Important Growing Countries.

Sl. No.	Name of the Country	Variety	Area (in Acres)	Produc-tion (in M.T's)	Exports (in M.T's)
1	Afghanistan	N.A.	N.A.	N.A.	N.A.
2	Bangladesh	N.A.	N.A.	N.A.	N.A.
3	Egypt	N.A.	N.A.	N.A.	N.A.
4	China	N.A.	N.A.	N.A.	N.A.
5	Pakistan	N.A.	N.A.	N.A.	N.A.
6	Iran	Malas Maykhosh Alok, taloubarik	1,00,000	6,00,000	20,000
7	Iraq	Ahmar. Aswad. Halwa	N.A.	N.A.	N.A.
8	India	Ganesh, Mridula, Bhagna	1,60,000 Anantput Hyderabad (A.P.) Bijapur, Chitradurga (karnataka) Sangola, Sholapur (Mahara-shtra)	7,00,000	5,000
9	Israel	Mule Head Wonderful	2,500	300	
10	Saudi Arabia	Mangulati	N.A.	N.A.	N.A.
12	Spain	Mollur deelchi Valcncianas	5,300 (Alicante Province	20,000	10,000
13	USA	Wonderful Graneda	3,500 (California)	17,500	No fresh produce export (except concentrated juice)

Information on other countries will be updated as and when the information is received from the contact sources. Pomegranate is not listed in FAO and other Statistical sources separately.

LIST OF CHEMICALS REQUIRE COMPULSORY TESTING FOR EXPORT

Sl. No.	Chemical	European union	UK	Dutch
Organo Chlorine Chemical				
1	Adrin	Banned	0.05	0.05
2	Chlordane Banned	0.02	0.02	0.02
3	Chloro thalonil	1.00	1.00	1.00
4	DDT	Banned	0.05	0.05
5	Dichloflunid	10.00	15.00	10.00
6	Dicotol	2.00	2.00	2.00
7	Doldrin	Banned	0.05	0.05
8	Endosulfan	0.50	0.50	0.50
9	Endrine	0.01	0.01	0.01
10	Lindane	0.50	0.01	0.01
11	SCH	=	=	0.01
12	Heptachlore	0.01	0.01	0.01
Organophosphorus				
13	Accephate	0.02	0.02	0.02
14	Aggeenphos methyl	1.0	2.0	1.00
15	Chlormotenivphos	0.05	0.05	0.05
16	Chloropyriphos	0.50	0.30	0.50
17	Chlorophyriphosmethyl	0.20	0.20	0.20
18	Dizinon	0.02	0.02	0.02
19	Dichlorovos	0.10	01.10	0.10
20	Dimethoate	1.0	0.02	0.02
21	Fthion	0.50	0.50	0.50
22	Etrimphos	=	=	0.05
23	Fenhlorophos	0.01	0.01	0.01
24	Fenhlorophos	0.01	0.01	0.01
25	Malathion	0.50	0.50	0.50
26	Profenophos	0.50	0.50	0.50
27	2-Chlorophenol	-	-	-
28	Metharnidphos	0.01	0. 01	0.01
29	Mthedethion	0.50	0.05	0.05
30	Mevinphos	0.10	0.10	0.10
31	Parathionethyl	Banned	=	0.05
32	Phosphonidon	0.15	=	0.15
33	Fosmite	Banned	=	0.05
34	Parathion methyl	0.20	0.20	0.20
35	Quinolphos	0.05	0.05	0.05
36	Trizophos	0.02	0.02	0.02

Cynthate parathroid				
37	Gyclothrin	0.30	0.30	0.30
38	Cyclothrin lamda	Banned	=	0.02
39	Cypermethrin	0.50	0.50	0.50
40	Deltamethrin	0.10	0.10	0.10
41	Fenvalarate 0.02	0.02	0.02	
42	Parmithrin	1.00	0.05	1.00
Trizins				
43.	Altrazin asayalannins	0.10	0.10	0.10
44	Banlexil	0.20	0.20	0.20
45	Metalaxil	2.0	2.0	2.0
Carbamate				
46	Carbofuron	0.10	0.10	0.10
47	Methomyl	0.05	0.05	0.05
Paramidins				
58	Fenarimal	0.30	0.30	0.30
Trizols				
59	Bitertenol	0.05	0.05	0.05
60	Fusiloczol	Banned	Banned	Banned
61	Hexaconozol	0.02	0.10	0.10
62	Propeconizol	0.50	0.50	0.50
Imidozoles				
63	Imizalil	0.02	0.02	0.02
64	Captofol	0.02	0.02	0,.02
65	Captan	3.00	3.00	3.00
Dicorboxy mides				
66	Imrodine	10.00	10.00	10.00
67	prosamidon	5.0	5.0	5.0
68	Vinchlozolin	5.0	5.0	5.0
Benzimizol				
69	Carbandizim	2.0	2.0	2.0
70	Thiophenate methyl	1.0	1.0	1.0
Thiocarbamate Fungicide				
71	Mancozeb	0.01	0.01	0.01
72	Maneb	2.00	2.00	2.00
73	Thiram	0.01	0.01	0.01
74	propinab	0.01	=	3.00
75	Lthylene Thiourca	0.01	=	0.02
Others				
76	Thiomethaxam Oxydemation methyl	@=0.02	@=0.02	@=0.02

77	Fostylal	#	#	0.20
78	Copper oxychloride	2.00	20.00	20.00
79	Tridemefon	-	2.00	2.00
80	Sulphur	50.00	=	50.00
81	Microbutanil	1.00	1.00	1.00
	Denocap	=	=	0.05
	Cymaczonil	0.01	=	0.05
	Carbaryl	3.00	3.00	3.00

As per Netherland MRI can be used @ syngenta Co. MRL, 0.1 mg kg, but further reduced to 0.02 kmg kg. (Dept of Horticulture. Maharashtra state) Pune-411005.

Annexure –III

Chemicals permitted for United Kingdom & European export markets

A)	Leaf spot & Fruit spot	Name of Insecticide	ppm	Days before harvest
1)	Bitratenanol	Baycor	2.00	14 days
2)	Bordeux Mixture	R.B.Bordo, Bordo, Bordo 20 W	5.00	07 days
3)	Captan 50% WP	Captaf, Captan, Mazcap, Herbicap, Captaf, Dhanutan, Kistal	2.50 to 3.00	50 days
4)	Chlorothalonil 75% WP	Karach, Jatayu	2.00	50 days
5)	Copper hydroxide 77% WP	Kocidelol,	2.50	07 days
6)	Copper oxychloride 55% WP	Blitox, Fytolon Blue copper, Copox, Danukap, Canupee, Blue diamond, Copter, Copper guard	2.50 to 3.00	07 days
7)	Difenoconozol 25% EC	Score	0.5 to 1.00	14 days
8)	Dodin 65% WP	Super Star, Noor, Sailit	1.00 to 2.00	14 days
9)	Fenarinal 12% EC	Rubigon	0.40	30 days
10)	Hexaconozol 5% EC	Contof, Sitara, Titan hexasole, Samarth, Conzol	1.50	14 days
11)	Iprobanepass 48% EC	Kitazin, Beltof	2.00	30 days
12)	Phytozyne Bycarbonate	Phytozyne Bycarbonate	2.50 to 5.00	07 days
13)	Potassium metabysulphide	Potassium metabysulphide	2.5	07 days
14)	Profalexin	Profalexin phosphoric acid	3.00	#
15)	Sulpher 80% WG/WP	Sultaf, Kosawet, Thiowet, Sulfex	1.00 to 2.50	10 days
16)	Tridenefan 25% WP	Beletan	1.00	04 days
17)	Tricyclezol 75% WP	Beem, Civic Tistar	0.80	#
18)	Captan 50% WP + Hexoconozole 5% EC	Takat	1.00	60 days
B)	**Bacterial leaf spot**	**Name of Insecticide**	**ppm before harvest**	
1)	Bromo-2, Nitropropane-1,3-Diol	Bactromycin plus, Bacterimycin 2000,	0.42	07 days
2)	Kasugamycin 3% L	Kasu B, Biomycin,	1.00	07 days
3)	Phosphoric acid	Phoscon, Ecomin, Icomin, Macimin, Decore	1.00	07 days
4)	Potassium bicarbonate	Potassium bicarbonate	2.50 to 5.00	07 days
5)	Streptomycin sulphate	Streptomycin sulphate	0.15	07 days
6)	Streptocyclin	Streptocyclin	0.40	07 days
7)	Rhizobacteria	P-suraksha, Pseudomonas, Ecomonos	2.00 to 3.00	07 days

C)	Fruit borer	Name of Insecticide	ppm	Days before harvest
1)	Alphamethrin 10%	Pharsa, Tataalpha, Stop, Sherpa, Alphaguard, Alphamark, Alphacure, Auxix, Alphin, Alphodhon, Gum	1.00	40 days
2)	Cypermethrin 25% EC	Bsathrin, Folithrin, Jawa, Cymbush, Cuper guard, Tata cyber, Kambush, Cybil, Silcord, Shakthi laser, Super lall, Indothin rolo	0.50	40 days
3)	Indoxcorb 14.5% EC	Dhaksh, Dhawa, Avanta, Rombo, Kingdoxa	0.50	30 days
4)	Metorizem and Beuaria (mixture)	Metabiv	1.00	#
5)	Novaludan 10% EC	Rena	1.00 to 3.00	25 days
6)	Sepinoside 45 EC	Taser, Simstar, Konserve	0.20 to 0.44	50 days
7)	Chlorpyripos Sypermethrin	Nural 0D,	2.00	40 days
8)	Thio-d-carb	Larvin	2.00	40 days
D)	**Mites**	**Name of Insecticide**	**ppm**	**Days before harvest**
1)	Abamactin 1.9% EC,	Vertinec	0.25 to 0.50	30 days
2)	Fenzakavin	Magister	2.00	28 days
3)	Millbemectin 1% EC	Millbecnaf	0.50	30 days
4)	Proparhite 57% EC	O.mite, Allmite	3.00	15 days
5)	Chlorophenpar	Interpred	2.00	15 days
6)	Decophole	Kurnel-S, Dicophot Kelthane	2.00	100 day
7)	Tridaymorph	Calixin	2.00	40 days
E)	**Thrips**	**Name of Insecticide**	**ppm**	**Days before harvest**
1)	Acetopromide	Record, Tata manic pride, Irban, Bhadsha Armer, Sabuka, Rapid, Guide, Bismark	0.10 to 0.20	30 days
2)	Azardiactin 10,000 ppm	Neem gold, Numbica plus Achuk, Neemark, Nimbicidin, Econeem plus, Nimozol – T/S	1.00	03 days
3)	Carbosulphon 25% EC	Electra, Audh, Marsh	1.50	30 days
4)	Deltamethrin 2.8% EC	Decis, Decaguard, Ecomethrin	1.00	30 days
5)	Fepronil	Regent	1.5 to 2.00	14 days

6)	Imediachloropid 17.8% SL	Confidor, Tatameda, Aldima, Champion chemide Tez, Zembo Emigrin, Imidasal, Hotshot, Confidant, Confidence 555	0.25 to 0.75	15 days
7)	Labadasahclothrin 5% EC	Karate, Laksh, Reeva-5, Halothrin	1.00	40 days
8)	Thiomethaxon 25% WG	Actara, Tataananth, Evident, Urga, Chick	0.20	15 days
9)	Accephate	Acetop	2.00	40 days
10)	Byfenthin	Tallstar	2.00	60 days
E)	**Mealy Bug**	**Name of Insecticide**	**ppm**	**Days before harvest**
1)	Azardactin 10,000 ppm	Neemark, Econeem plus, Nimbica plus, Neem gold, Nimbicidin, Nimizol – T/S	1.00	03 days
2)	Cal MB	Cal MB	2.00	03 days
3)	Verticoliem lacani	Biosar, Dharsha	1.00	03 days

Chemicals not permitted for United Kingdom & Europian export markets

A)	**Leaf spot & Fruit spot**	**Name of Insecticide**	**ppm**	**Days before harvest**
1)	Carbandizim	Bavistin, Dersole, Bengard, Benfil	1.00	90 days
2)	Tridaymorph 80% EC	Calixin	0.50	40 days
3)	Mancozeb 75% WP	Indofil M-45, Bilzeb, Dithanium-45,	2.50	70 days
4)	Metriram 70% EC	Polyram	3.00	90 days
5)	Propiconizol 25% EC	Tilt	0.25 to 0.50	30 days
6)	Propineb 70% WP	Antracol	3.00	90 days
7)	Zyneb 75% WP	Indofil-Z-78	2.50	90 days
8)	Ziram 27% S:L	Cuman-L, Zyrux-L	3.00	90 days
9)	Benomil 50% WP	Benomil	0.30	90 days
10)	Metalexil Mancozeb	Ridomil	1.00	66 days
11)	Fosetyl AL 80%	Allittee	1.00	90 days
12)	Fenconozol	Topaz	2.00	50 days

B)	Fruit Borer	Name of Insecticide	ppm	Days before harvest
1)	Carbaril 50% WP	Devicarb Carbane, Sevin Moxicarb, Sulphoril	2.00	50 days
2)	Dichlorovos 76% EC	Nuvan, Luhan, Masivos, Doom-Dash, Demand, Domer	1.00	30 days
3)	Endosulphan 35% EC	Indotof, Indosole, Thiodan, Thiokill, Sulphan	2.00	30 days
4)	Quinolphos 25% EC	Ekalux, Splash, Kinolkhus, Berocil	1.25 to 2.00	20 days
5)	Chloropheriphose-Methyle	Dursban, chloroguard, REdar, Reldan	1.00 to 2.00	30 days
6)	Ethophan	Etherel, Diphal, Ethipal,	1.00	100 days
C)	Mites	Name of Insecticide	ppm	Days before harvest
1)	Ethion 50% EC	Mit 505, Dhanumit 50, Nilmite, Afathenion, Mitkill Lazor, Fosmite, mit	2.00	100 days
D)	Thrips and Aphids	Name of Insecticide	ppm	Days before harvest
1)	Dimethoate	Rogar, Nuvagore, Didhan, Tara Milgore	1.50 to 2.00	20 days
2)	Mithyleporethion 50% EC	Milion, Metacid, Dhanumar, Bukti Metapar parasul	3.00	24 days
3)	Monocrotophos 36% EC	Tatamono, Monocil, Cotocil, Monosan, Liuphos, Nuvocron	1.50	50 days
4)	Oxidematon methyle 25% EC	Metasystox Highmax	2.00 to 2.50	24 days
E)	Mealybug	Name of Insecticide	ppm	Days before harvest
1)	Dichlorovos	Nuvan, Doom Demand, Deevap, Doomer	1.00 to 2.00	30 days
2)	Melathion 50% EC	Melathion, Nagmola	2.00	40 days
3)	Methomil 40% SP	Lanate, Dunate, Dash, Astra	1.00	30 days

Color Plates

Kesar (Bhagawa, Shendria, Ashtgandha)

Ganesh (GBG)

Phule Arakta

Ruby (IIHR)

Single Stem

Multi Stem

Bacterial blight

Brown to dark brown necrotic spots on the fruits
Red to pinkish pigmentation around the necrotic area brown pigmentation

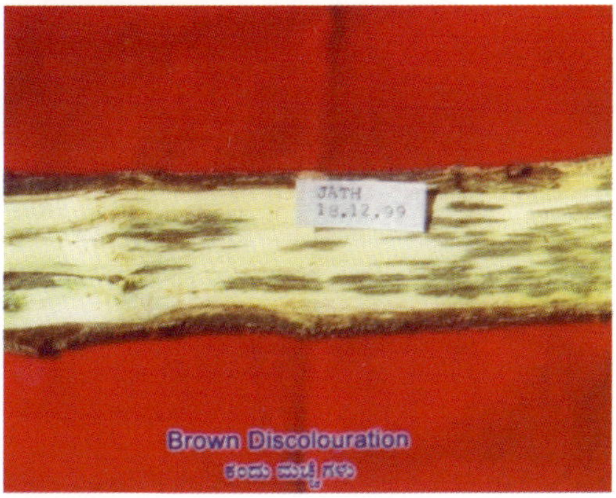

Wilt affected plant

Yellow halo around the necrotic lesions

Advanced stages- Big necrotic lesions on the leaves

Symptoms on affected twigs

Initial infection on twigs

Bluish discoloration on the young tender shoots

Rough and cankerous lesions on twig

Rough and cankerous lesions on main stem

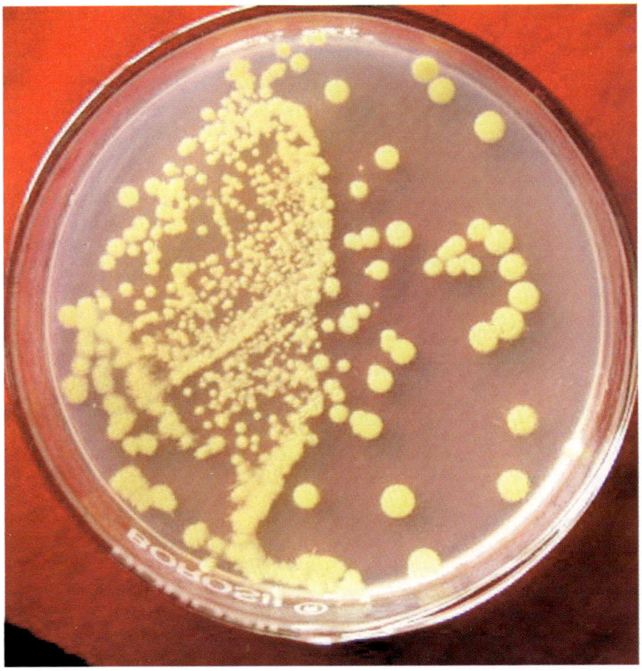

Colony character of ***Xanthomonas axonopodis*** pv. *punicae*

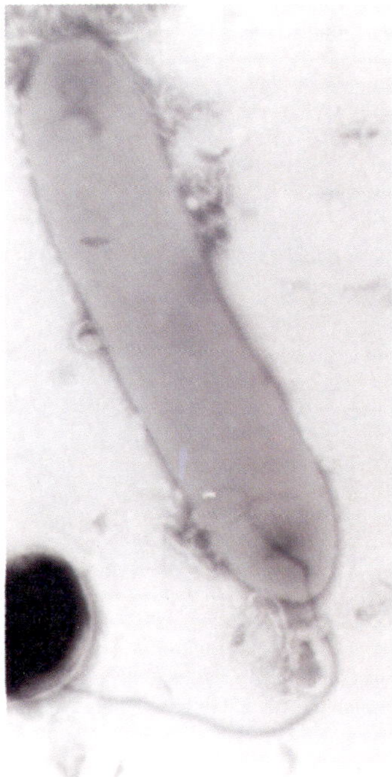

Electron microscopic picture of *Xanthomonas axonopodis* pv. *punicae*

Enumeration G-ve bacterial cells trapped on glass slides

Glass slides fixed at different heights above the ground level (feet)	G-ve cells/microscopic field									
	With in the garden					Distance from the source of infection				
	Block 1	Block 2	Block 3	Block 4	Block 5	0.5 km	1.0 km	1.5 km	2.0 km	2.5 km
1.5	0.00	0.00	0.00	0.00	0.00	0.00	0.00	0.00	0.00	0.00
3.0	2.33	4.99	6.66	5.33	9.00	10.66	8.00	7.33	0.00	0.00
4.5	10.33	13.33	25.00	19.33	11.33	27.00	23.66	14.00	36.00	0.00
6.0	12.66	30.00	43.33	37.33	19.00	41.33	33.33	29.33	21.00	0.00
SEm±	1.2802	1.6159	2.6771	2.0207	1.6915	0.5528	0.3727	0.5528	0.4082	
CD@ 1%	6.0741	7.669	12.7018	9.5877	8.0256	2.6227	1.7682	2.6227	1.9370	

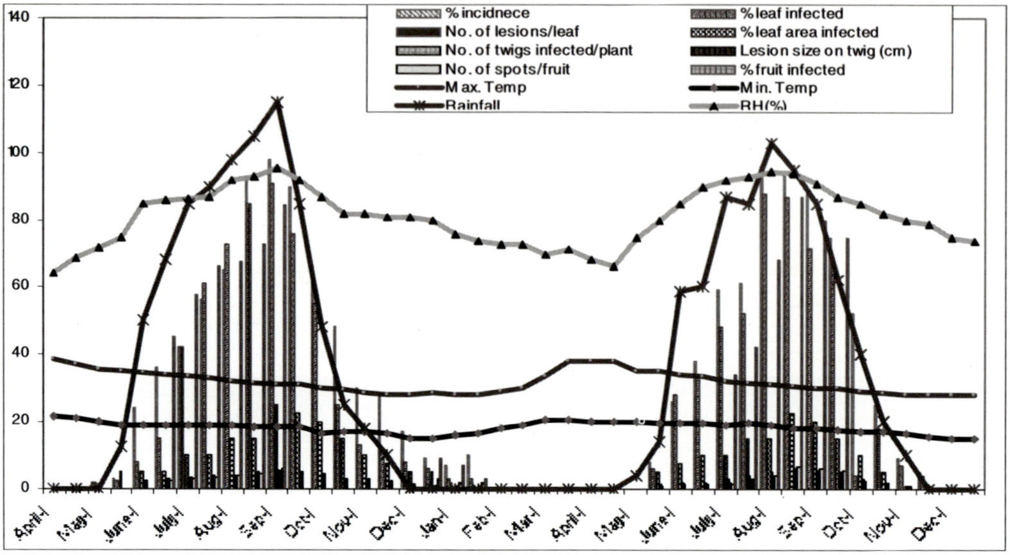

(Courtesy: Dr. Nazeer Ahmed Khan Rtd. Professor of Plant Pathology, UAS, Bangalore)

Anthracnose

Bacterial leafspot

Fruit sucking moth damage fruit

Anthracnose

Phomopsis

Bacterial leafspot (Oily)

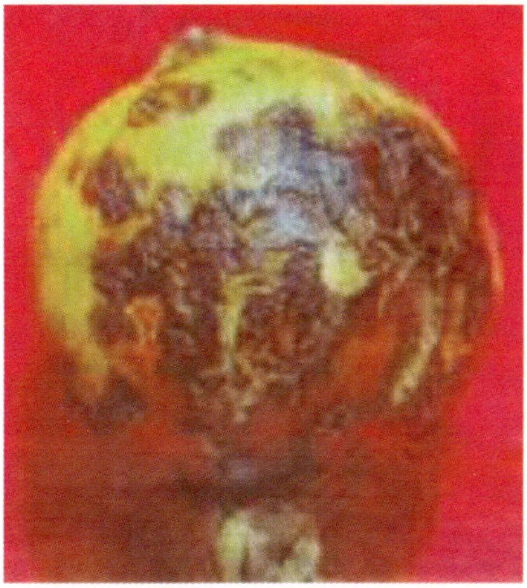

Bacterial blight

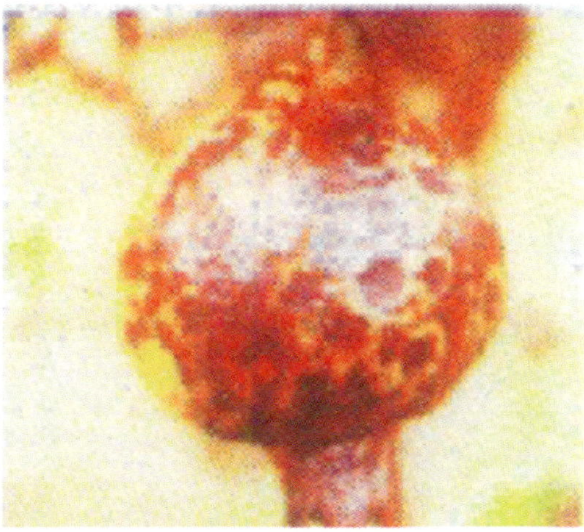

Anthracnose

Alter naria

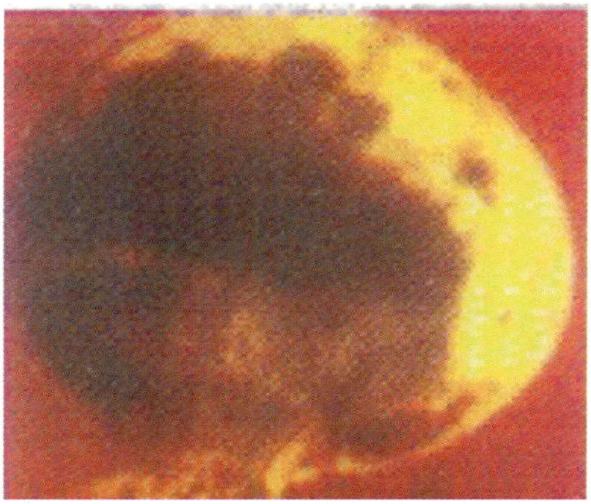

Phytopthora

Cercospora

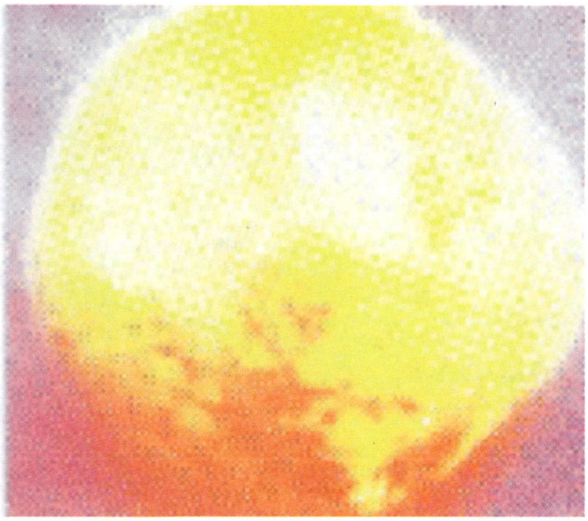

Scab

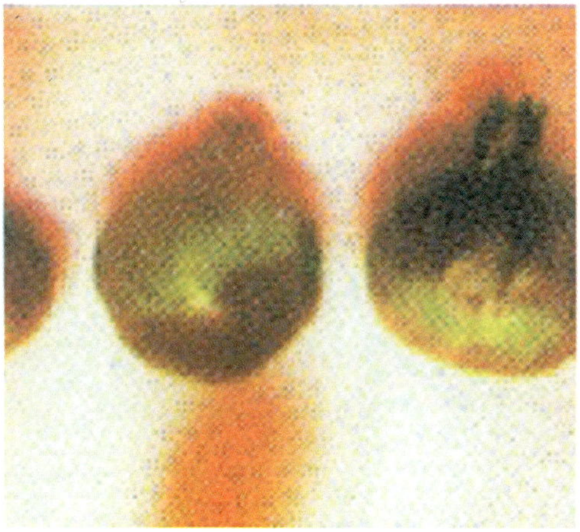

Phomopsis

Die back

Index

P

R

S

T

U

V

W

Z